BIOCHEMICAL EDUCATION

Biochemical Education

Edited by
CHARLES F. A. BRYCE

CROOM HELM LONDON

© 1981 Charles F. A. Bryce
Softcover reprint of the hardcover 1st edition 1981
Croom Helm Ltd, 2–10 St John's Road, London SW11

British Library Cataloguing Publication Data

Biochemical education.
   1. Biological chemistry – Study and teaching
   I. Bryce, Charles F A
   574.1'92'07

   ISBN-13: 978-94-011-6552-5   e-ISBN-13: 978-94-011-6550-1
   DOI: 10.1007/978-94-011-6550-1

Typeset by Pat Murphy IBM, Highcliffe, Dorset

Biddles Ltd, Guildford and King's Lynn

To Maureen, Christopher and Jonathan
for having a lot of patience and understanding

CONTENTS

CONTENTS

PREFACE

The purpose of the present text is to distil the experience of a number of workers active in the field of biochemical education, so providing readable accounts which, it is hoped, will be of significant benefit to those who are new to the teaching profession in addition to those who may be stimulated to experiment with alternative strategies in their own teaching situation.

From the various contributions considered in this book, each topic, in its widest sense, would warrant at least a volume on its own and indeed such texts are currently available. However, it was felt more appropriate to restrict the coverage to those aspects which are of particular use to the subject of biochemistry and, for which, work in this area has already achieved some measure of success. In effect what each of us is doing is supplying findings from a body of knowledge collectively called educational technology. Without entering the debate on the semantics of what educational technology is or is not, it doesn't take long to realise that, like the vast majority of subject areas, it has its own unique terminologies and vocabulary. Whilst it is inevitable that such terms will appear throughout the text, hopefully all will be explained on first use and so it is not envisaged that this will be too distractive to the reader. The rationale behind this type of approach is that if learning can be facilitated and improved, while at the same time motivating the student and increasing the interest level in the subject, then this is an exceedingly profitable exercise.

I would like to express my gratitude to the contributing authors for their time and effort in preparing their respective chapters of this book. With regard to the preparation of the manuscript, I am indebted to Janice Guthrie for producing the figures and tables, to George L. Hunter for the photographic work and to Anne Carrie for an excellent job of typing the manuscript and, in the process, correcting some spelling mistakes which had found their way into the draft manuscripts.

1 THE SYSTEMS APPROACH IN BIOCHEMISTRY

Charles F. A. Bryce

Introduction

The overall aim of this book is to promote an active interest in the process of facilitating and improving student learning and in the planning of the learning process. A structured protocol for such an aim is called a learning system and the mechanism for achieving it has been termed the systems approach (Gagné, 1962; Churchman, 1968; Carter, 1969; Davis *et al.*, 1974). The purpose of this first chapter is to provide an overview of a systems approach to learning to act as a backdrop against which the following chapters can be perceived and to highlight the interrelationships which exist between the various topics discussed in later chapters. A scheme of the systems approach is shown in Figure 1.1.

Figure 1.1: A Schematic Representation of the Systems Approach to Learning, Illustrating the Interactions which Exist and the Hierarchical Structure

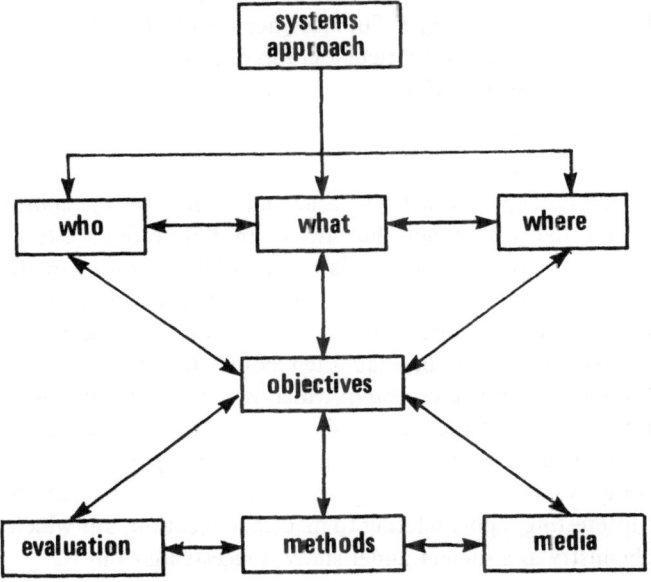

The 'Who'

The 'who' deals with such aspects as the age, number, academic status
and expectancies of the students involved and such a group is termed
the target population. The teachers can also be considered under this
heading from the point of view of number available, particular
expertise, etc.

A good example of this type of planning can be seen in the reports
by Treagust and Cody who, in 1975, participated in the design of the
Secondary Science Training Program for Biochemistry (Treagust and
Cody, 1977; Cody and Treagust, 1977). This program was developed
for *highly-motivated* secondary-school students entering junior or
senior years, who would be required to live in the university
dormitories during the summer vacation in order to have ready access
to the library, laboratories and the summer seminar program. Another
easily-identifiable group considered under the heading 'who', which
requires selective consideration, would be medical or nursing students.
Here the particular emphasis is on providing a biochemistry course
which is both comprehensive and academically acceptable and one
which is relevant to, and integrates with, their clinical studies (Saffran,
1971; Rubinstein, 1972; Spilman and Spilman, 1975; Stohs and
Rosenberg, 1976; Hultquist *et al.*, 1976; MacQueen *et al.*, 1976). The
considerations applicable to this particular group of students are dealt
with in some detail in the following three chapters.

The 'What'

The 'what' section is concerned with the academic content of the course,
the subject matter, the syllabus and the curriculum. With reference to
biochemistry, and in particular the very fast rate of information growth
in this subject, there is an obvious problem in deciding what material
should be included in an undergraduate syllabus and what should be
excluded. All too often there is an established 'core' which is deemed
fundamental to the subject and on top of which is developed a number
of important or more recent topic areas (Fox, 1978). The danger then
is to include more factual information than can reasonably be
assimilated in the available time and the net result is one of information
overload. An interesting report relating to medical education, but which
includes biochemistry as a subject under study, suggests that this is
indeed the case and that we are probably trying to put over about two
to four times too much factual material (Anderson and Graham,
1980).

The 'Where'

The 'where' deals with somewhat disparate considerations such as the physical location of the learning (for example small tutorial room, large lecture theatre, field trips, laboratory, computational laboratory), the equipment that would be necessary and the time allocation for the teaching session.

The Objectives

The objectives include the general aims of the course, the specific objectives and taxonomies for these and also considerations based on the distinction between cognitive, psychomotor and affective domains of achievement.

Traditional syllabi normally appear in print to highlight the cognitive learning that is desired, with in some cases, a more detailed analysis in terms of specific learning objectives. If, however, one were to ask the lecturers involved, many would suggest that psychomotor skills were an important consideration and others would argue that attitudinal aspects were equally important. Thus, for some courses the development of interpersonal skills is a component part of the course (Hall, J., 1973; Tribe and Peacock, 1973) in others the development of value judgements is seen as important (Long, 1979). Such considerations are dealt with in more detail in Chapter 6 whilst the assessment of learning in the affective domain is discussed in Chapter 5. Some useful guidelines relating to topic analysis and programme planning are discussed in a recent article on cell biology by Fedoroff and Opel (1978).

The Media

This heading refers to the selection of an appropriate medium from a wide choice of single and mixed-media formats including tape, tape/slide, film, tape/film, television, print and computer-assisted learning, and would also include such things as budgeting considerations and availability of specialist equipment.

It has been stressed that such technological hardware cannot supplant a human being (Charren, 1980), that there is nothing magic about them and that their effectiveness lies in what a good teacher does with them (Freitag, 1980). Another concern in this area is one of simple logistics in that some teachers have so much trouble arranging for the use of this specialist equipment and, in addition, it is often unreliably maintained so that they find it easier just to stick to chalk and talk (Brandt, 1980).

The Methods

The methods refer to the actual way in which the learning is to be achieved, for example by lecture, small-group learning or problem-solving, and also involves such considerations as safety aspects and whether or not the method is suited to the nature of the task being considered.

Here, as at other points in the systems approach, the ideal selection is not always feasible because of particular constraints. For some educational establishments a fairly common constraint results from the human resources, particularly when there are reasonably heavy teaching commitments. In a recent report, one biochemistry teacher solved the problem of having a heavy teaching load and also a desire to be actively involved in research — he combined the two activities and found that the students benefited from this development (Jones, 1976). Another common constraint, part of a group collectively termed 'institutional constraints', refers to the limitations or lack of specialist accommodation and equipment. For example, imagine the difficulties faced with trying to demonstrate to a class of students the experimental procedure involved in protein sequence analysis using ultracentrifugation, amino-acid analysis, dansylation of the intact protein, tryptic and chymo-tryptic cleavage with peptide fractionation and characterisation! A solution to overcoming the obvious constraints involved in this exercise was to use photographs and facsimilies of the experimental data in order to allow the student the opportunity of analysing critically the data for himself (Leader, 1976). Other examples of this approach are given in later chapters.

The Evaluation

This final heading deals with the assessment of the learning outcome and may involve a choice of assessment method, the use of scoring keys or impression marking, criterion-referenced or norm-referenced scoring, and reliability and validity of the assessment method (Ashworth, 1972; Callely and Hughes, 1972; Harper, 1972; Thomas, 1972; Varley, 1972; Billing, 1973). Assessment is plainly an essential part of the educational process — it serves as an index of the effectiveness of the teaching, it provides a mean for the classification of student performance and it helps to provide the student with detailed feedback on his/her own performance.

Hierarchy of Systems Approach

Figure 1.1. highlights the interactions between the various elements of
the system whilst at the same time indicating the hierarchical nature of
the approach. Thus strictly speaking, one should normally consider a
new problem first in terms of the who, what and where. Once this has
been achieved, the objectives are spelt out and these joint considerations
would then provide some guidelines on which to select the appropriate
media, methods and evaluation for that particular learning unit
(Popham and Baker, 1970; Davies, 1971; Corrigan, 1980). I say
'normally' because some workers do not actually go through each stage
in an established well-defined order but rather 'find' their way through
the systems approach having given consideration to each of the main
headings (Cowan, 1980; Wildman, 1980). This flexibility which is a
feature of a mutually-interacting whole can be considered an advantage
in achieving the optimal solution for a particular problem. Having said
that, the corollary does not hold, in other words one should not start
off in designing a learning package by saying 'I'd like to produce a
half-hour colour TV programme on enzyme kinetics!'

The remainder of this chapter provides a number of specific
examples of the way in which a systems approach has been applied in
biochemistry. Each of the learning packages discussed will be considered,
where appropriate, in terms of four distinct phases – design, production,
implementation and evaluation. Following this, the later chapters will
then go on to deal with different aspects of such an approach in
significantly more detail.

Topic 1: Chemical Composition of Nucleic Acids, Nucleotides and Related Compounds

Design Phase

Target Population. The student groups for which these packages have
been produced are given below:

(1) BSc Science, Years 1, 2
(2) BSc Nursing, Years 1, 2
(3) HND (Higher National Diploma) Biology, Year 2
(4) HNC (Higher National Certificate) Biology, Year 2
(5) HNC Medical Laboratory Technicians, Year 2.

The students, male and female, average age 18–21, have in most cases an educational background of at least one year of a biological science course in which the fundamentals of biochemistry are outlined as a brief overview.

Topic Selection. For the courses selected there are a number of topics which are common to all of the syllabi. One such topic is the chemical composition of nucleic acids, nucleotides and related compounds and this routinely appears in traditional syllabi as 'DNA Structure', 'Chemical Composition of Nucleic Acids', 'The Structure of Nucleotides and Nucleic Acids' etc.

Here, as with all other topics dealt with, it is advisable before starting on the design of the package to make sure that no one has already done the work, otherwise it is duplicated effort. If the work has been done before, either as published in a journal, commercially available or provided by a colleague, then the chances are that it will not be 100 per cent satisfactory for the particular needs of the course being considered and so will need some modifications before being implemented. Having said that, however, such modifications can often be very trivial matters and, with much of the spade-work being done already, this can be a very profitable avenue to take.

Time Allocation. The material to be dealt with should be the equivalent of two lecture-hours of instruction.

Aims and Relevance. One of the aims of this learning package is to demonstrate that, despite their apparent complexity, nucleotides and nucleic acids are relatively simple molecules to study both in terms of biological structure and function. It is generally accepted that one of the main problems in teaching biochemistry to undergraduate students is the structural complexity of the molecules involved. To generate such structures visually in the lecture is either very time-consuming or inappropriate. The solution is to use either pre-prepared materials (slides, models, overhead transparencies, text etc.) or to present the structures in a highly simplified way as shown in Figure 1.2.

Figure 1.2: Schematic Representation of a Nucleic Acid

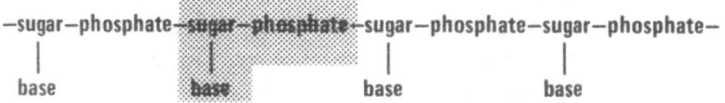

Clearly the latter is more useful for the student under test from the point of view that he/she can easily reproduce simple line drawings and discuss these. It is arguable, however, that such an approach may lead to an oversimplification of what is really present at the molecular level and so prevent/inhibit fundamental understanding of the subject.

Terminal Objectives. By the end of the learning package it is envisaged that each student should, without reference to his or her notes or textbook, be able to:

(1) define the terms base, nucleoside, nucleotide, oligonucleotide, polynucleotide, phosphodiester bond, pyrophosphate bond and cyclic nucleotide;
(2) outline the chemical constituents of both DNA and RNA;
(3) describe the relationship of a nucleotide to a nucleic acid;
(4) distinguish between a purine and a pyrimidine ring system;
(5) state which bases found in DNA and RNA are derived from purine and which from pyrimidine;
(6) discuss what is meant by the term *minor base*;
(7) number correctly the sugar moieties according to the accepted convention;
(8) draw the structures for each of the major bases;
(9) draw the structures for ribose and deoxyribose;
(10) draw the structures for a given nucleoside, nucleotide and oligonucleotide;
(11) distinguish between mono-, di-, tri-, and tetra-nucleoside phosphates.

Given in this way, the students themselves are able to identify just exactly what they are required to be able to do after using the learning package. In the initial stages of this work our students were largely unfamiliar with being given the objectives of the course (or part of it) in this way but, from a number of informal discussions with them over a period of about five years, the concensus is very definitely in favour of their use.

The purpose of using learning objectives in this way is to highlight, both for the staff and students alike, just precisely what is required from the course. Thus, while the syllabus indicates the topic areas, it does not give any indication of the level and standard of the desired outcome. A lecturer's reaction, when he is introduced to such learning objectives for the first time, is probably to say that they are trivial and too detailed;

that certainly was my own reaction. However, if one overcomes this recalcitrance then a more lengthy exposure can lead to the view that the use of learning objectives can be a very worthwhile process indeed. One of the hidden benefits in the construction and use of objectives in this way is that it actually makes the lecturer stop and think carefully as to just exactly what he does want the student to learn and be able to use, and this in itself is not an insignificant bonus. There is a vast literature concerned with the pros and cons of the use of learning objectives (Broudy, 1970; MacDonald-Ross, 1973; Davies, 1976; Calder, 1980) and whilst this material can be very helpful in moulding one's own opinion, the only way for an individual to arrive at a purposeful judgement as to their potential is to try constructing and using them in his own teaching environment.

What then is a learning objective? It is in essence a statement of what the student should be able to do at the end of the teaching process and is confined to one limited outcome. It should be precisely written and contain three essential elements: (1) an observable action, (2) conditions under which this action will occur, and (3) a minimal acceptable standard of performance (McAshen, 1970). This can best be illustrated using a specific example from a workbook on the IUB Enzyme Classification System (Bryce, 1975):

> By the end of the prepared lecture notes, each student should, without reference to his or her notes be able to:
> — list in the correct sequence the six main classes of enzymes.

In this example, the observable action is that the student *lists six main classes*, the conditions under which this should be done are *without reference to his or her notes* and *by the end of the prepared lecture notes* and lastly, the minimal acceptable standard is that the list is in the *correct sequence*. In some cases the standard can be omitted and in these cases it is an implied standard of 100 per cent. The requirement for the action verb to be an observable action is related to the assessment procedures and also to the behaviourist theory. It is argued that in order for the student or teacher to know if material has been assimilated then the student must demonstrate this in a tangible way. This requirement excludes the use of verbs like *understands* or *appreciates* since these processes cannot be observed, although some workers in this area would argue strongly to the contrary (Wight, 1972; Thomas, 1976). From the theoretical aspects of educational technology applicable to the learning process, we find that there are different levels

of student attainment ranging from knowledge (Bloom, 1956) or
stimulus-response (Gagné, 1962) at the low levels, to evaluation
(Bloom) or problem-solving (Gagné) at the highest level as summarised
in Figure 1.3. For a more detailed account, the reader is referred to the
original texts.

**Figure 1.3: Taxonomies of Educational Objectives of the Cognitive
Domain**

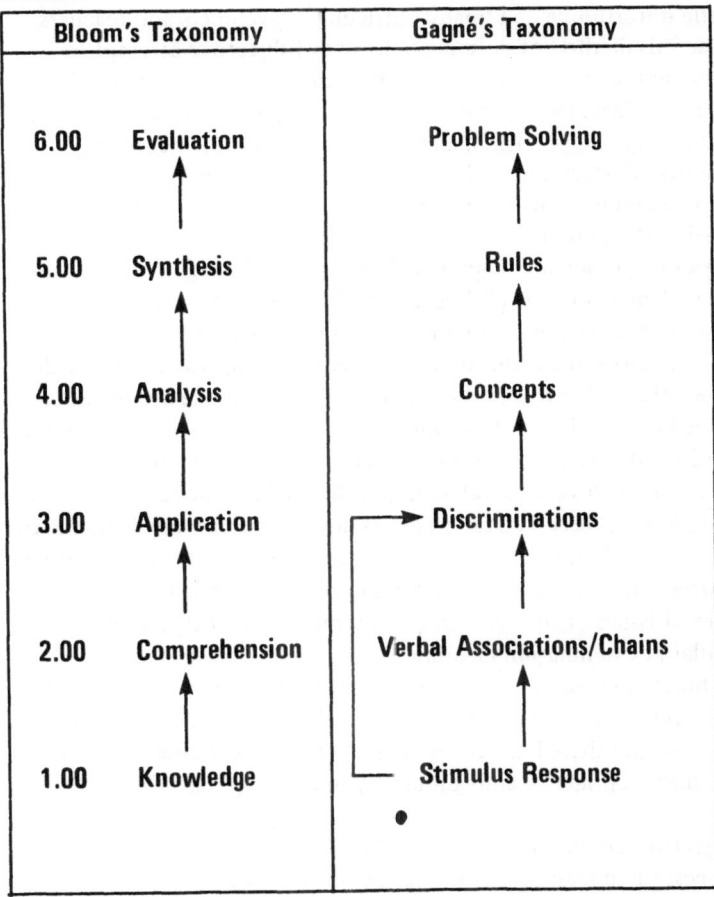

It has been stated that the use of learning objectives is ideal for
those items of simple recall but quite inappropriate for defining and
assessing higher-order skills and indeed their use can result in the over-
emphasis of the latter (Popham and Baker, 1970; MacDonald-Ross,

1972). This need not necessarily be the case as has been shown recently by a number of workers (de Winter Hebron, 1979; Race, 1979; Huxham and Naeraa, 1980). Examples of the way in which learning objectives have been used in biochemistry can be seen in the reports by Maffet (1967), Beard and Pole (1971), Rasmussen (1972), Beard (1973), Tribe *et al.* (1975), MacQueen *et al.* (1976) and Davies (1980).

In many references to objectives they have been termed 'behavioural objectives'. As Davies (1976) points out, the connotation *behavioural* is a little unfortunate and 'many curriculum developers and teachers who use this method of defining objectives [objectives attempt to describe, in the clearest possible terms, exactly what a student will think, act or feel at the end of the learning experience], do so for the sake of clarity and precision, and not because they are necessarily behaviourists'. Davies in his book uses the term specific objectives as an alternative and it is this term and also the term learning objective which are used in the present chapter.

Davies begins a later chapter with the quote 'All argument is against it; but all belief is for it' [Johnson] and he goes on to discuss fairly extensively the arguments that have been put forward for and against the use of objectives in an educational context. My own experience with using learning objectives in the way described for the learning packages contained in this chapter has convinced me that, in my own educational establishment, the case for using objectives is strong. They have provided me with very useful guides on the syllabus material and this information was used at a later date in developing new curricula for two courses in which the syllabi are written completely in terms of learning objectives. Their use has also helped in peer communication on a number of issues of the syllabus (or interpretation of it) and in particular at examination meetings.

As mentioned earlier, the students involved in the packages described in this report have, without exception, welcomed the use of objectives in this way and there has been no complaint that the objectives are trivial, inappropriate or ambiguous. Again to paraphrase Davies:

> Objectives are not abstractions. They represent commitments to the process of education. . . . We define objectives so that we can say what we mean, so that we can explicitly determine what it is that we wish to do. . . . Defining objectives often brings into sharp focus the underlying questions of relevancy and validity . . . and is merely one facet of the much more complex process of planning.

Topic Analysis

To be in any way effective, students using this material are assumed to have a certain level of prerequisite knowledge which defines their entry behaviour. Thus, before progressing on to this learning package, each student should, without reference to his or her notes or textbook, be able to: (1) identify the chemical structure of a sugar, (2) draw the structure of a phosphate group, and (3) distinguish between a simple molecule and a macromolecule.

Learning Hierarchy. Once the learning objectives have been constructed each should be compared with every other one in order to determine if there are any relationships between them. Such an analysis should then facilitate the logical construction of what is known as a learning hierarchy. Learning objectives themselves can be classified by a number of taxonomies (see earlier) and range from low-level cognitive behaviour (Bloom, 1956; Gagné, 1962) and similar sub-classes for the affective domain (Krathwohl *et al.*, 1964) and the psychomotor domain (Harrow, 1972). A number of workers suggest that, in ordering the learning objectives to form a hierarchy, the low-level objectives should form the base of the hierarchy. Once these have been mastered the student progresses through successively more demanding objectives until he/she completes the package. An alternative view is held by Bruner (1966) who argues that the learner should first master the most complex objectives and then proceed to the more basic by discovery. The findings of Bruner were later developed into the Concept Attainment Model of Joyce and Weil (1972). In this area of educational technology one very quickly becomes confronted by a long list of models to help explain information processing, the inductive and deductive models, the Taba model, the Ausubel model, the Suchman model and so on (Eggen *et al.*, 1979). Whilst a number of workers in science education have derived valuable aid from consideration of such learning models, it is true to say that most were derived from studies dealing with the teaching of young children. This being the case, a number of workers have recently questioned any attempt to extrapolate from pedagogy to androgogy (Cronin, 1979; Stewart, 1979). For the present topic the former approach of structuring the objectives was adopted and a scheme for the learning hierarchy is shown in Figure 1.4.

Sequencing Sub-topics. When dealing with a system ranging from small molecules to large macromolecules, the most logical way to treat this is to start with the most simple and go through the intermediate structures

Figure 1.4: A Learning Hierarchy for a Teaching Package on Nucleic Acids, Nucleotides and Related Compounds

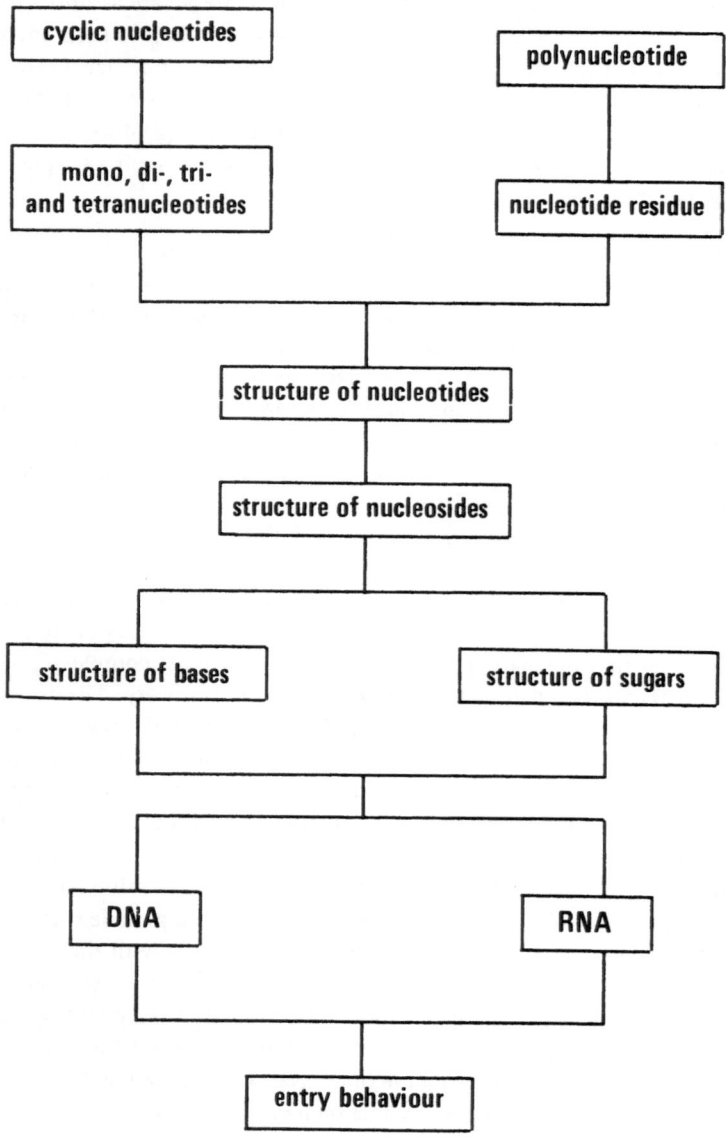

continuing in complexity until we arrive at the macromolecule. At the same time we go from the general to the specific as shown schematically in Figure 1.5.

Figure 1.5: Sequencing Sub-topics for a Teaching Package on Nucleic Acids, Nucleotides and Related Compounds

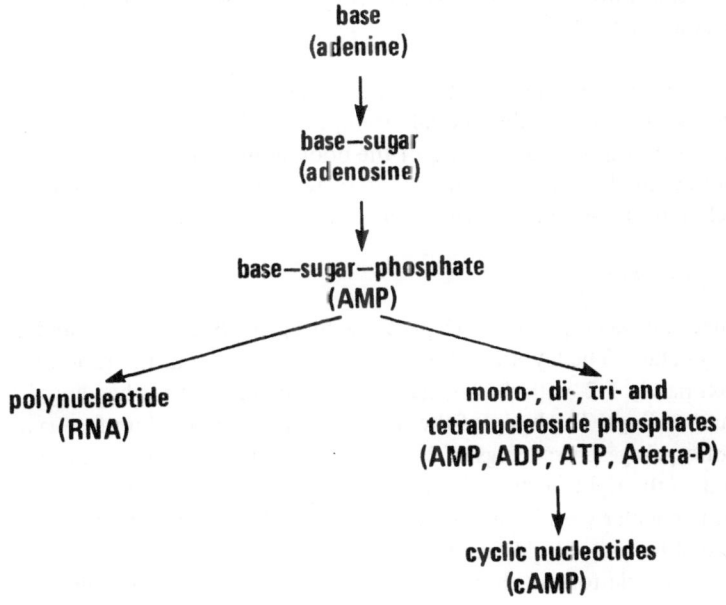

Media and Methods Selection. The medium selected for this package was print. The reasons for this choice were as follows:

(1) Production costs are relatively inexpensive.
(2) the content does not require any associated complex visual material, animation, colour cueing and would not, it was felt, benefit from an audio component and for these reasons a number of other media/mixed media could be disregarded.
(3) With the majority of the new material being chemical structures, which different people come to terms with over different times, any medium which was externally paced, for example tape-slide, PIP or video, would be inappropriate. In view of the sometimes large differences in student ability within classes and between classes, a package that was self-paced was essential. This lack

of ability need not be taken as something sinister, but instead, may merely represent a 'slow gelling time' for that particular part of the syllabus.

To date, the material has been presented to the students in one of two ways, either as front-line teaching in which case the student was left to do the work himself, or it was used in conjunction with formal lectures on this material.

Student Evaluation. The workbook was designed to contain objective test items which the student could attempt. Detailed explanatory answers were supplied at the end of the booklet in order that the student could check his/her own progress, realise any misconceptions and relate progress to the learning objectives of the package.

Production Phase

The material was typeset on A5 page-size using an IBM composer and a 10 pt typeface. The layout of the print was off-centred in the ratio of approximately 1:2, with the left-hand section being used for headings and subheadings. The body of the text was typeset in medium-face type with interparagraph spacings of 2–4 picas. Headings and subheadings (ranged to the right of the left section) were typed in 10 pt bold-face. In this particular workbook, key words within the body of the text were cued by use of bold-face type.

The rationale for the choice of a particular layout, type-size, line length, typographical cueing etc. was based on the recent empirical findings of a number of workers actively researching this area of communication (as discussed in considerably more detail in Chapter 8). A copy of a page from the workbook is reproduced to illustrate the points described (see Figure 1.6).

Validation Phase

When the workbook was first produced, its effectiveness was assessed by my colleague, Dr David Button of the Department of Molecular and Life Sciences. In working through the package he was considering a number of criteria:

(1) Were there typographical errors?
(2) Were there errors in the scientific content?
(3) Were there any significant omissions of material?

Figure 1.6: An Example of the Typographical Layout of a Specimen Workbook

minor bases

In discussing the bases found naturally in DNA and RNA we were careful in the last section to say that in either DNA or RNA we "**principally find**" four different bases.

The reason for this is that in addition to these bases we also have a large range of so-called **minor bases.** These are called 'minor' because they occur less frequently than the others.

Some of these minor bases will be discussed more fully in later sections of nucleic acid biochemistry. It is sufficient at this stage of learning to know that such compounds can and do exist naturally.

sugars

In nature, there are only two types of sugar present in nucleic acids, **ribose** which is present solely in RNA (hence its name) and **deoxyribose** which is present solely in DNA (again the sugar gives rise to the name **deoxyribo**nucleic acid).

The chemical structures for these compounds are shown below:

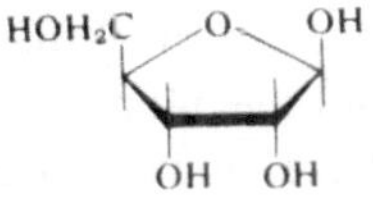

ribose

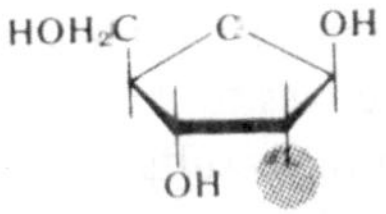

deoxyribose

Again you are expected to know these structures and be able to distinguish between the two. The prefix '**de-oxy**' means '**without – oxygen**' and we

6

(4) Was any statement ambiguous or misleading?

(5) Were any of the test items included in the workbook inappropriate?

(6) Were there any alterations to the layout which would improve the visual presentation?

Fortunately in this case, the answers to most of the above questions were negative and only two typographical errors required correction. The reason 'for getting it right' first time is largely attributable to two main factors. First, the material is fairly standard and in no way is it currently a controversial issue (not the case with some later topics). For this reason the same material is acceptable to all the teachers involved in its delivery. Secondly, having taught this material on several occasions to a number of classes over a period of years, I was able to pinpoint any problem areas and highlight the most relevant material so much more easily as a result of these 'dry runs'.

The next phase in its evaluation was to use it with a class of students and note any problem areas in its use. With some of the groups, the students were, without warning, given a pre-test, consisting of 20 objective test items, on this material about a month before they studied the nucleic acid section of their syllabus. Copies of the workbook (with the typographical errors corrected) were supplied to the students who were then given two weeks to study the material after which time they were given a formal objective test containing 20 different test items on the material. With one class (BSc Science, Year 2) the students were also given a post-test without warning about one month after completion of this part of their syllabus. The results are shown schematically in Figure 1.7.

Clearly from the results, the teaching package was effective in the teaching of the specific learning objectives to the students. Further, there was no significant difference in the results between the use of the package as (1) a self-learning unit or (2) in association with corresponding lecture material.

Topic 2: Simulations of the Physical and Chemical Properties of the Nucelic Acids

Design Phase

Target Population. For some of the learning material considered in this

Figure 1.7: Results of Pre-test Scores and Post-test Scores for a Single Class of Undergraduate Students

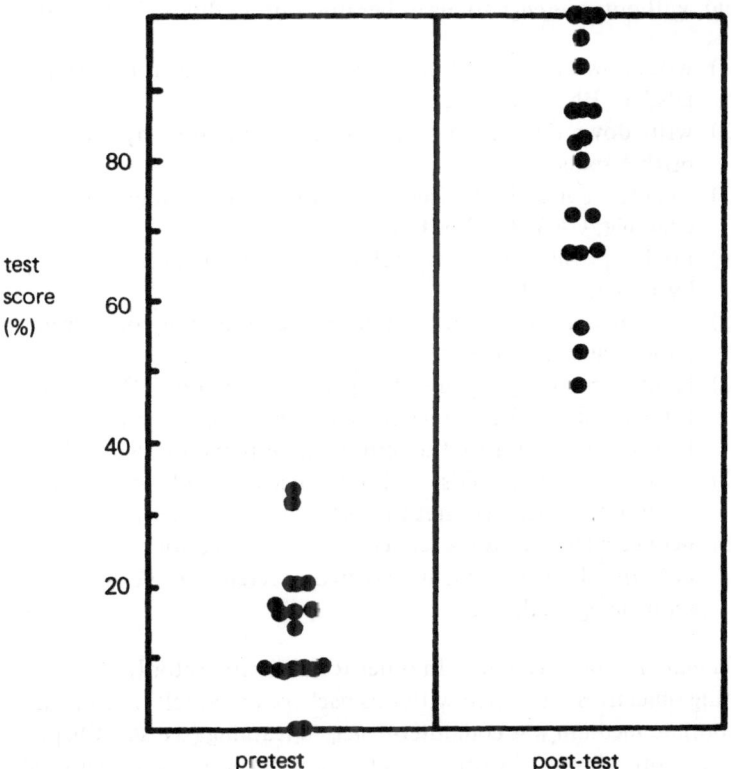

topic, the target population was students in the first year of their biochemistry course; other parts of the material were more appropriate for the second-year biochemistry course and for the remainder, the material was best-suited for the Honours-year course (fourth-year students having studied the subject of biochemistry for three years). Each of the sub-topics was in fact developed independently but, for the purposes of this chapter, are brought together as a single unit.

Topic Selection. The subject material that represents the main topics includes such issues as the structural aspects of nucleic acids, nearest-neighbour frequency analysis as an experimental technique and rapid nucleic-acid sequencing methods.

Structural Aspects of the Nucleic Acids

Terminal Objectives. By the end of the learning package, each student should, without reference to his or her notes or textbook, be able to:

(1) write down the correct complementary sequence for a given DNA or RNA sequence;

(2) write down the correct translational product for a hypothetical mRNA sequence;

(3) calculate correctly the frequencies of specific codons, for example, UAA, AUG or UGA;

(4) predict possible secondary folding arrangements for a hypothetical mRNA;

(5) evaluate correctly the extent of secondary folding for a given molecular arrangement;

(6) identify correctly regions of a specific nature on a DNA strand, for example A—T rich region, palindrome regions, Pribnow or Hogness box, putative transcriptional termination signal;

(7) identify and characterise with reference to a table of enzyme specificities, cleavage sites for restriction endonucleases;

(8) deduce a DNA or RNA sequence from gel-electrophoretic patterns obtained from at least two different nucleic-acid sequencing methods.

Media and Methods Selection. In order to fulfil satisfactorily the learning objectives associated with this package it was felt that the most appropriate medium was computer-managed learning (see also Chapter 9). The approach was to write a simple computer program which would generate a large number of random polymer sequences (either DNA, RNA or RNA with minor bases). A certain number of these, usually six, would be given to each student along with an associated questionnaire from which they would carry out a number of discrete exercises. Clearly, the emphasis/rationale of such an approach is very much drill and practice, a strategy which appears well-suited to meet the specific learning objectives.

For the purposes of this learning package it was felt that there was no requirement for making use of the computer in an interactive mode. Indeed there were many advantages to be gained by having the source material generated in a batch mode by the lecturer, since this obviated the necessity of booking a computer laboratory and it allowed the package to be given to the students as a home exercise which could be submitted in parts rather than as a whole.

Student Evaluation. For each exercise the student was required to submit a completed questionnaire, which had to be structured by the lecturers involved, along with the copy of the random sequences which had been provided. The exercises were given a test score plus any appropriate comments and returned to the student as feedback.

Production Phase

The computer program which generates the random polymer sequences, POLSEQ, is written in BASIC and occupies a file of three pages of store (one page being equivalent to 512 36-bit words). For the lecturer to obtain batch output of either DNA, RNA, RNA with minor bases or protein/peptide sequences merely requires running the program with any one of four specific data statements present. These clearly can be altered very easily from one run to the next. Alternatively, the program can, without modification, be operated by the use of one of four control files. These preclude the user from having to alter this DATA statement. In addition, these short programs direct the output to the line printer and this is often very much more convenient. It was an easy matter to modify and adapt the computer program to output peptide/ protein sequences which would also allow learning materials to be produced in the area of protein sequence analysis. The program POLSEQ has also been implemented on a number of different micro-computers and desk-top calculators.

Questionnaires dealing with the aspects of RNA structure, DNA structure and protein/peptide structure were constructed on the basis of the learning objectives associated with these curriculum topics. These have taken a number of different forms over the last few years and can easily be adapted to suit the needs of a particular educational establishment or individual. The questionnaires were typeset on an IBM composer and A4 copies run-off as required. Copies of the current questionnaires are reproduced in Figures 1.8, 1.9 and 1.10.

Apart from the advantage of providing the student with drill and practice in the area of polymer chemistry, this type of study allows the student to be introduced to certain material which might otherwise be ignored or glossed over. For example, the exercise on predicting RNA secondary structures serves as an excellent introduction to the work of Fresco, Alberts and Doty (1960) and has spin-off when considering tRNA structure, 5S RNA and 16S RNA structures. In the same vein, consideration of the frequency of cleavage sites for restriction endonucleases in relation to their specificity is a worthy discussion point.

Figure 1.8: Specimen Student Questionnaire Related to the Topic of the Structure of RNA

RNA Sequences

Complete the following exercises using the random-generated sequences provided. Return the completed questionnaire with the associated computer listing.

1. For each of the sequences, evaluate the base composition and record the results in the following table.

Base	Frequency					
A						
U						
G						
C						

2. Assume that the first sequence represents a short stretch of mRNA, write down the product formed as a result of translation of this mRNA.

3. Do any of the 6 sequences have the initiation signal as the first three bases on the 5' end?

yes	no

    If 'yes', which sequence?

4. Starting from the left hand end of each of the sequences and progressing along them in triplet codons, can you find one termination codon in the correct phase?

yes	no

    If 'yes', which sequence and where in the sequence does this occur?

sequence number	base position

Figure 1.8 − *continued*

5. Based on the example shown below, choose any one sequence and predict a possible
    secondary folding arrangement.   Draw your folded structure in the grid supplied.

AAUCUCCGGUAAGAUGGUGCCUGAUUCUCACUCACGCCGCAAAGUCGAGU

Predicted secondary structure for a random-generated RNA sequence containing 50 bases. The resultant
structure contains a number of hydrogen-bonded regions connected by loops of different size and composition.
From the base composition of the sequence, the secondary structure outlined represents one in which 75%, of the
total possible number of base pair combinations are formed

6. Count the number of base pairs in your predicted structure (shown above) and determine
    the extent of secondary folding (expressed as a percentage of the maximum possible).

number of base pairs	maximum number of base pairs	% secondary folding

Figure 1.9: Specimen Student Questionnaire Related to the Topic of the Structure of DNA

DNA Sequences

Complete the following exercises using the random-generated sequences provided.   Return the completed questionnaire with the associated computer listing.

1.   For each of the sequences, evaluate the base composition and record the results in the following table:

Base	Frequency					
A						
T						
G						
C						

2.   Write down the sequence of the complementary strand of sequence 1.

3.   For the first ten bases of sequence 2, write down the complementary strand sequence and include the number of hydrogen bonds between each base pair.

4.   Assume that one of these DNA molecules undergoes transcription to yield a corresponding mRNA.   Does the initiation codon (AUG) appear in phase in any of these mRNA molecules?

yes	no

If 'yes', which sequence and where in the sequence does this occur?

sequence number	base position

Figure 1.9 – *continued*

5. Do any of your sequences have AT-rich stretches (i e. more than 8 consecutive bases)?

yes	no

If 'yes', which sequence and where in the sequence does this occur?

sequence number	base position

6. Do any of your sequences have palindrome regions?

yes	no

If 'yes', shade in the region with a coloured pencil/pen.

7. Given that the specificities of certain restriction endonucleases are as shown below.

restriction endonuclease	specificity
Alu I	$AG{\downarrow}CT$
Bam H-I	$G{\downarrow}G\,ATCC$
Eco RI	$G{\downarrow}AATTC$
Hae III	$GG{\downarrow}CC$
Hha I	$GCG{\downarrow}C$
Hind II	$GTPy{\downarrow}PuAC$
Hind III	$A{\downarrow}AGCTT$
Hpa I	$GTT{\downarrow}AAC$
Hpa II	$C{\downarrow}CGG$
Pst I	$CTGCA{\downarrow}G$
Sal I	$G{\downarrow}TCGAC$

Would any of your sequences be cleaved by any of these enzymes?

yes	no

If 'yes', which enzyme, which sequence and where in the sequence does this occur?

sequence number	base position	restriction enzyme

Fig. 1.9 – *continued*

8. Do any of the 6 sequences contain within them the putative Pribnow box (or Hogness box in the case of eukaryotes)?

yes	no

If 'yes', draw a box in the appropriate position.

9. Is there any evidence for the putative transcription termination signal AATAAA?

yes	no

If 'yes', underline the appropriate sequence.

10. Using sequence 4 as a template in DNA sequencing experiments, draw the electrophoretic band pattern which you would expect to find by (i) the Maxam and Gilbert method and (ii) the 'plus-and-minus' method of Sanger and Coulson.

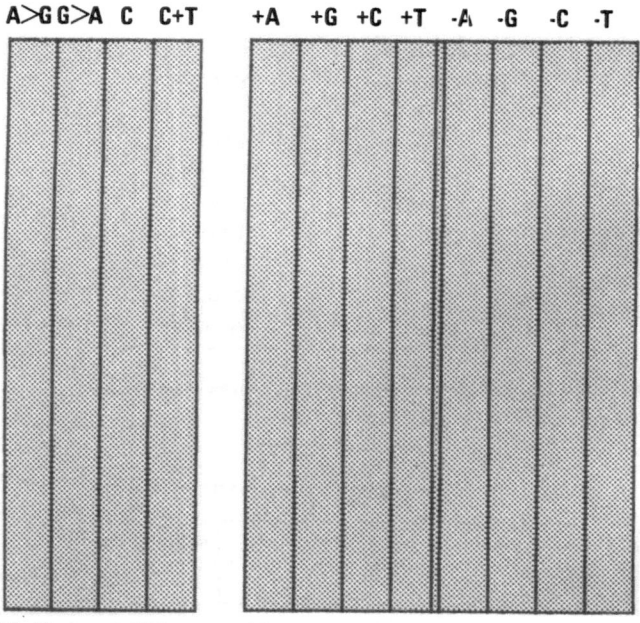

(i) Maxam & Gilbert   (ii) Sanger and Coulson

Figure 1.10: Specimen Student Questionnaire Related to the Topic of the Structure of Proteins and Peptides

Protein/Peptide Sequences

Complete the following exercises using the random-generated sequences provided. Return the completed questionnaire with the associated computer listing.

1. Complete the following amino acid analysis table for sequence 1.

amino acid residue	no. of residues
cysteine	
aspartic acid	
threonine	
serine	
glutamic acid	
proline	
glycine	
alanine	
valine	
methionine	
isoleucine	
leucine	
tyrosine	
phenylalanine	
histidine	
lysine	
arginine	
tryptophan	

2. How many of the amino acid residues shown in this sequence are hydrophobic in character?

Shade these residues in using a coloured pen/pencil.

3. How many of the amino acid residues shown in this sequence are hydrophilic in character?

Shade these residues in using a different coloured pen/pencil.

Figure 1.10 – *continued*

4. Given the following table of dissociation constants for particular amino acid residues, calculate the pI of peptide sequence 2 at pH 7.0.

amino acid residue	pKa
arginine	12
ε-lysine	10
tyrosine	10
histidine	7
aspartic acid	4
glutamic acid	4

Your estimate for the pI value is

5   Would the peptide discussed in question 4 migrate to the anode or cathode on electrophoresis at pH 7.0?

anode	cathode

6. Indicate on sequences 3 and 4 those areas/locations which you might predict would not participate in a stretch of α-helix

7. Indicate on sequence 5 those sites on the peptide which would be cleaved by treatment with trypsin.

8. Indicate on the same sequence those sites on the peptide which would be cleaved by treatment with chymotrypsin.

9. Which sequence would you expect to possess the highest molar extinction coefficient at 280 nm?

Why?

This package has been implemented by other members of staff, both in my own college and in a number of other educational establishments, and no further changes seem to be required in its implementation as a result of a summative evaluation.

Design Phase

Nearest-neighbour Frequency Analysis. In many undergraduate biochemistry courses there is a section of the syllabus on the mechanism of DNA replication and the enzymes involved. This raises the question for the student 'How do we know that the DNA is being faithfully copied from the template rather than just a random DNA-like polymer being produced?' There are a number of pieces of evidence which the lecturer can discuss in this context, but the definitive answer to the question comes from a study of the technique of nearest-neighbour frequency analysis (NNFA).

For a number of reasons, undergraduate students appear to find great difficulty in understanding the basic rationale of this technique. Whilst accepting that a number of these misconceptions can be easily discussed and resolved, the traditional approach to discussing such a technique, namely the lecture, suffers from the fact that normally a single set of experimental data is discussed and analysed without actually giving the student the opportunity of sitting down with his own unique set of data and making similar conclusions. Likewise, within a class of students, it was often apparent that different students met different problems in understanding the technique. As a result, these two observations suggested some form of self-paced individualised learning packages as a possible solution to these problems (Bryce, 1978).

Terminal Objectives. By the end of the learning package, each student should, without reference to his or her notes or textbook, be able to:

(1) describe correctly the experimental protocol of NNFA;
(2) distinguish correctly between the specificities of a number of nuclease enzymes;
(3) list at least four distinct properties of DNA for which NNFA can act as a diagnostic tool;
(4) list the materials essential for *in vitro* DNA-dependent DNA biosynthesis;
(5) explain how to interpret nearest-neighbour frequencies in terms of:
  (i) the polarity of the strands

(ii) the base equivalence
(iii) random or template-directed
(iv) the fidelity of replication
(v) whether all 16 possible dinucleotide sequences occur.

Materials and Methods. As described earlier, the most appropriate medium format for a particular package was only considered once the target population was specified and the learning objectives, topic analysis and learning hierarchy were determined. Such an approach was implemented for the package on NNFA and the most appropriate medium was identified as a computer-assisted learning format.

Production Phase

The computer program for simulating nearest-neighbour frequency analysis, NNFSIM, is written in BASIC and occupies eleven pages of disc store. The program itself is composed of three distinct phases, an introduction with instructions to the students on how to proceed etc., an assessment section to test the students' prior knowledge of the technique of NNFA (results of the assessment are output on the user's terminal at the end of this phase) and finally a simulation of a laboratory experiment which provides the student with a unique set of results on which to carry out a detailed analysis (see Figure 1.11).

One aspect of this study of which we were particularly conscious was the visual appearance of the text material on the VDU screen. We have made every effort in designing the program to make sure that material is both legible and well laid-out on the screen (see also Chapter 8). For example, when a question is posed in the assessment section then only that question appears on the screen. It was our opinion that, for example, remnants of a previous question and/or additional unnecessary text were distractive. Also, to prevent material at times 'running-off' the screen before the user has a chance to read and digest the information, we have used a small software sub-routine which serves to set the page size. For many users there is a systems command which has this very function (on the DEC–20 this is @ TERMINAL PAGE 20) but in practice these tend to be less reliable than one 'built' into the program itself. Such a sub-routine can be usefully used for a certain number of other functions within the program (Bryce and Stewart, 1979).

In the present version of the program the random-number generator statements are preceded by the statement RANDOMISE. The purpose of this is that since RND is a pseudo-random number generator function

Figure 1.11: Flowchart of the Program for Simulating Nearest-neighbour Frequency Analysis, NNFSIM

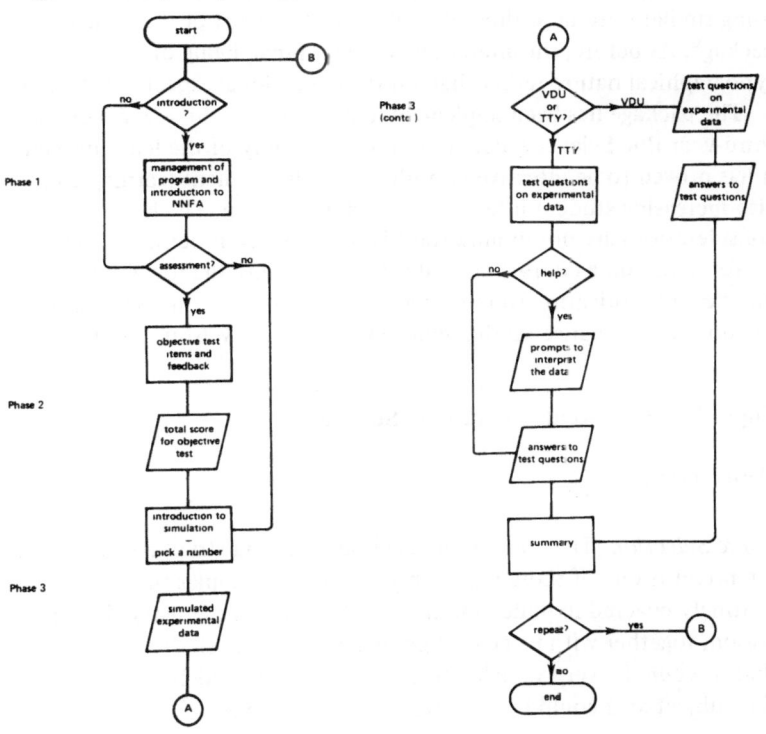

Source: Bryce, C. F. A. (1978) *J. Biol. Education,* **12,** 133.

then we get the same number in the same order each time the program is run. For many simulations or early versions of a new program, then there is an advantage in having this reproducibility. However, for the present programme it is more appropriate to have a truly random number. RANDOMISE has the property that it resets the numbers in a random way. For those computers which do not have such a facility an alternative approach is to have the student pick a number (say between 1 and 50) and use this value to cycle round the RND statement that number of times (Bryce, 1978) or to link it in some way with a clock facility on the computer or microprocessor.

There is also a BASIC-PLUS-2 version of the program NNFSIM which occupies 23 pages of disc store. The advantage of this version is that it can be stored in an executable form and so saves on compilation time.

Validation Phase

This learning material was initially assessed, again by Dr David Button, using similar criteria to those described in the report of the earlier package. As before, the alterations were minimal, being of typographical nature rather than on the educational aspects of the unit.

The package has been implemented over the last three years by our third-year BSc Science students and, from a study of the learning gain, it has proved to be effective in both facilitating student learning and also increasing student interest level and motivation. The latter considerations are not unimportant issues when considering student performance on a course and, indeed, in one study the low student morale and motivation were recognised as the major contributions to the poor performance on that course (Horsey and Milson, 1980).

Topic 3: Amino Acids and Protein Structure

Design Phase

Topic Selection. The area, amino acids and protein structure, like Topic 1 represents one of those reasonably well-defined topics that is routinely covered in undergraduate teaching in biochemistry. For this reason, together with its central importance to later studies, we felt that it would be very valuable to consider ways in which the teaching of this subject area could be improved and strengthened.

Time Allocation. The material subsumed under this heading would normally cover about six lecture hours of curriculum material.

Terminal Objectives. By the end of the lecture course and associated tape-film sequence, each student should, without reference to his or her notes or textbook be able to:

(1) describe the general properties of proteins as a class of compounds;
(2) discuss briefly the importance of proteins in life systems;
(3) identify the basic building-block in protein structures;
(4) identify correctly the 20 common amino acids contained in proteins and draw their structures in full;
(5) state the relationship between the properties of a protein and its constituent amino acids;

(6) describe how such basic building-blocks condense to give initially a dipeptide and eventually a polypeptide;

(7) discuss the important structural elements of the peptide bond;

(8) recognise and differentiate between primary, secondary, tertiary and quaternary structures;

(9) indicate by an appropriate diagram the relationship between these four levels of structural organisation;

(10) list and evaluate the advantages that can be accrued by a protein having a quaternary structure;

(11) summarise the structural characteristics of proteins, selecting appropriate examples to illustrate the points discussed.

Media Selection. Experience of teaching this material by formal lectures over a period of years has reinforced the view that there is often a very wide variation in students' abilities within a class for a particular topic. This need not necessarily represent the same variation based on overall academic status, but rather result from difficulty met by an individual student at a particular part of the syllabus. In short, this type of observation has been used in the past as a good case for individualising student learning. Without acceding to this global conclusion, we did, in fact, in the present study decide to select a media-format which was capable of achieving this purpose of individualised learning. In addition, we required a media-format which would allow us to generate fairly complex visual material, for example a line drawing of the α-helix, by a process of gradual build-up. For these reasons, the Philips PIP tape/film system was chosen, the equivalent equipment in the United States being the Besseler Cue and See.

The PIP system — *p*rogrammed *i*ndividual *p*resentation — is an audio-visual unit which combines both still and motion picture capabilities into a single and yet simple system. The visual component is contained within a film cassette which holds a maximum of 50 feet of unmodified Super 8 mm film, this corresponding to the equivalent of approximately 3,500 separate film frames whilst the audio-cassette is the standard cassette form. The viewing unit, shown in Figure 1.12, plays both sound and 8 mm film together and in synchronisation and this is achieved by electronic pulses on track 4 of the audio component (tracks 1 and 2 carry narration etc.) These pulses are inaudible to the user of the system and are recorded on to the tape using a special pulse generator (see Figure 1.12).

With this device it is possible to record pulses either for use in single frame changes or for use in animated sequences ranging from 1 frame

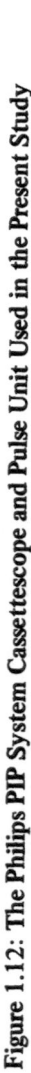

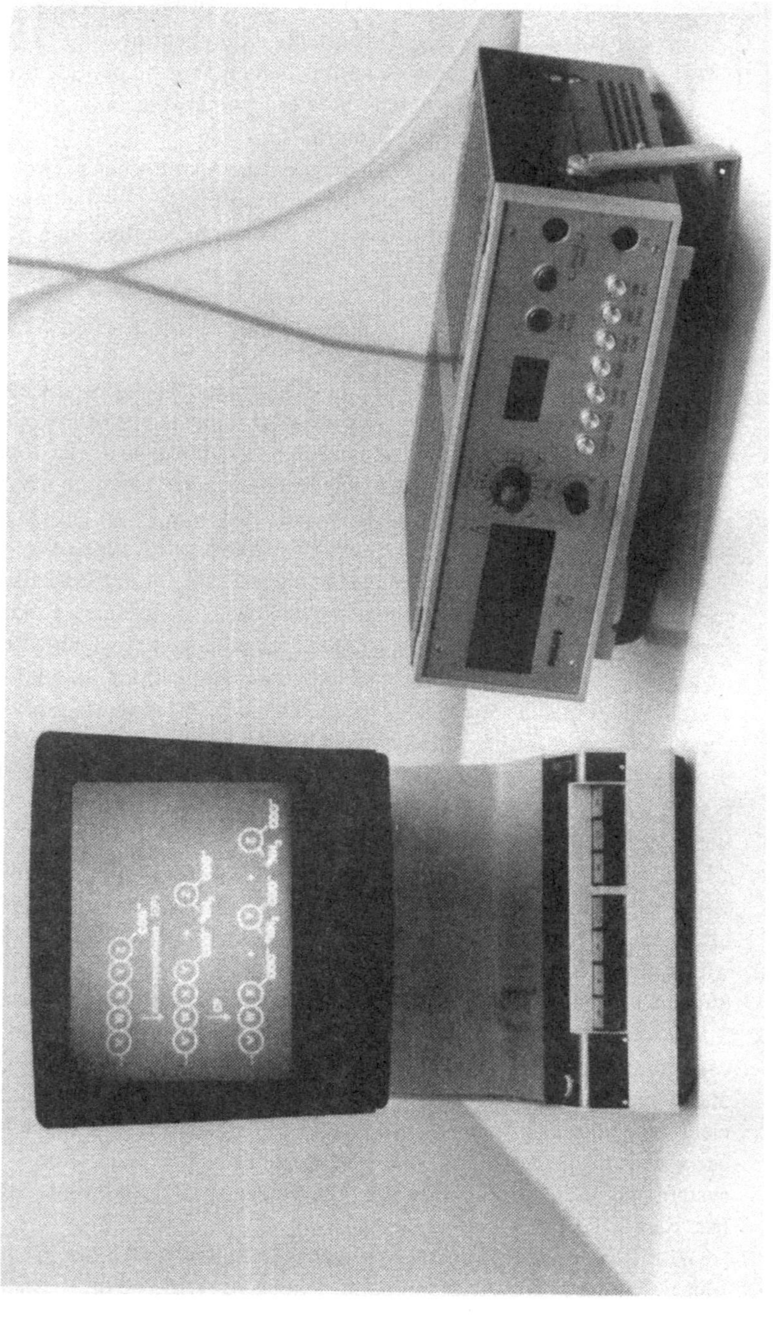

Figure 1.12: The Philips PIP System Cassettescope and Pulse Unit Used in the Present Study

per second (fps) or less to 24 fps which is the normal running speed of 8 mm film. The system is also designed to make use of these pulses as a 'skip facility' where, by suitable pulsing of the audio component it is possible to skip over any unwanted visual material. This facility allows the user of the system to design a single unit of software in terms of multi-level programming, multi-level both in terms of student level (i.e. good student, poorer student, revision purposes) and also subject level (fundamental, intermediate, advanced) (Bryce and Stewart, 1978). The rationale of multi-level programming is illustrated in Figure 1.13.

Figure 1.13: Multi-level Programming Using the Philips PIP System

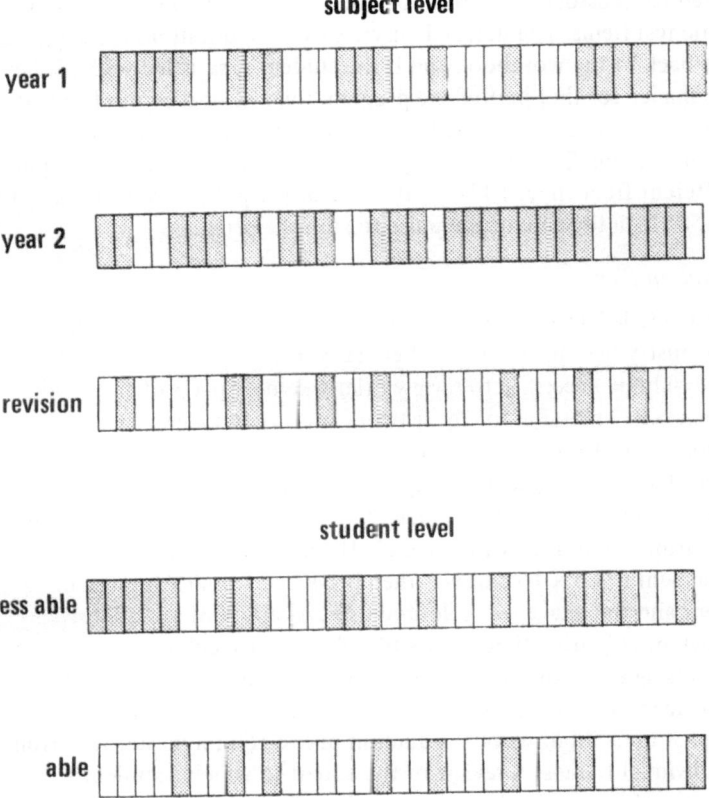

Before leaving this section it is worthwhile to point out that the facility for the gradual build-up of a frame need not be restricted to visual/graphic information but can equally well include textual information. The great advantage in this context is that because the visual information is synchronised exactly with the audio-tape there is no tendency to 'read ahead' of the commentary as is the case with, for example, tape-slide sequences. The net result is that the learning process is not inhibited by this type of distractive feature and plainly this is a significant and valuable characteristic to be gained from this media-format.

Student Evaluation. The learning packages are designed to stop at about five points during the program run-time. At these points the student is referred to an associated workbook and asked to attempt a number of specific test items. The detailed answers to these questions are supplied at the back of the workbook and if the student is satisfied with his/her performance for that part of the program then he or she goes forward to the next section; otherwise he can rewind the tapes and repeat the previous section. This stop facility is accomplished by the use of a pulse of different frequency, 1 kHz in the case of the pulse to advance a frame and 150 Hz in the case of the pulse to stop the program.

Production Phase

Since a detailed account of the production of PIP programmes in biochemistry has appeared elsewhere (Bryce and Stewart, 1978) it should only be necessary to review briefly some aspects of this work.

The textual material for the artwork was typeset using an IBM composer whilst the remainder was produced by a graphics artist at the college. To facilitate the next stage the conversion of these positive images from paper to negative images on film, the artwork was, as far as was possible, contained within a grid, 10 cm x 8 cm. A lith negative enlargement (21.5 x 16 cm) was then produced from each piece of artwork and the image was coloured using a variety of self-adhesive transparent colouring films (AGA filmolux, twelve colours). The reason for choosing a negative image polarity was initially subjective, but our experience to date has demonstrated that this mode is particularly beneficial in facilitating the production of the 16 mm master film from the individual frames. A review of the extent to which a student's learning performance is dependent on aspects of the physical environment, in particular the standard of projection, has recently appeared in the education literature (Wilkinson, 1976). With respect to whether one should use positive or negative images then, objective comparisons of the two are scarce and, unfortunately, frequently contradictory (Rubin, 1975; Nelson, 1965; Judisch, 1968; Block, 1968; Lee and Buck, 1975; Desrosiers, 1976) One problem which we encountered was a wide variation in the light transmission properties of the different coloured films. To circumvent this problem we used additional layers of certain colours in order that the final frame should have a mix of colours all of about the same density of light. This also provides the opportunity to blend layers of different colours to give additional shades to the original twelve colours.

The camera assembly used in the production of the PIP software is shown in Figure 1.14. It can be seen that the artwork is carefully positioned on a screen which is back-projected and, once positioned, single-shot filming of each image was carried out. Masking and unmasking of areas of a particular frame could be very easily achieved in this approach by the addition or removal of self-adhesive black masking film. The product from this work was a 16 mm master which was commercially converted to 8 mm copies by reduction printing, the latter being fixed into the plastic film-cassette cases. It is worthwhile pointing out that once a number of different programs have been written the various masters (all of which are still only relatively short lengths of film) can be spliced together, so reducing the actual production costs markedly. The audio component was recorded on a master tape which was, after editing, pulsed using the pulse generator. The pulsed master tape was then dubbed down on to four standard audio-cassettes using a high-speed copier.

Implementation Phase

At the present point in time we have 15 PIP units installed in the learning resources area of the college library for student use at any time during normal library opening hours. In addition there are two units installed in a small tutorial room adjacent to the biochemistry laboratory. For some classes the PIP material represents true resource-based material or front-line teaching and therefore these students are encouraged to make full use of the units. For other classes, the PIP programs represent additional related material which supplements and reinforces the lecture.

Validation Phase

All of these programs have undergone fairly extensive field trials and an analysis of the results indicates that the use of such teaching packages either in conjunction with the lecture or as front-line teaching is to be

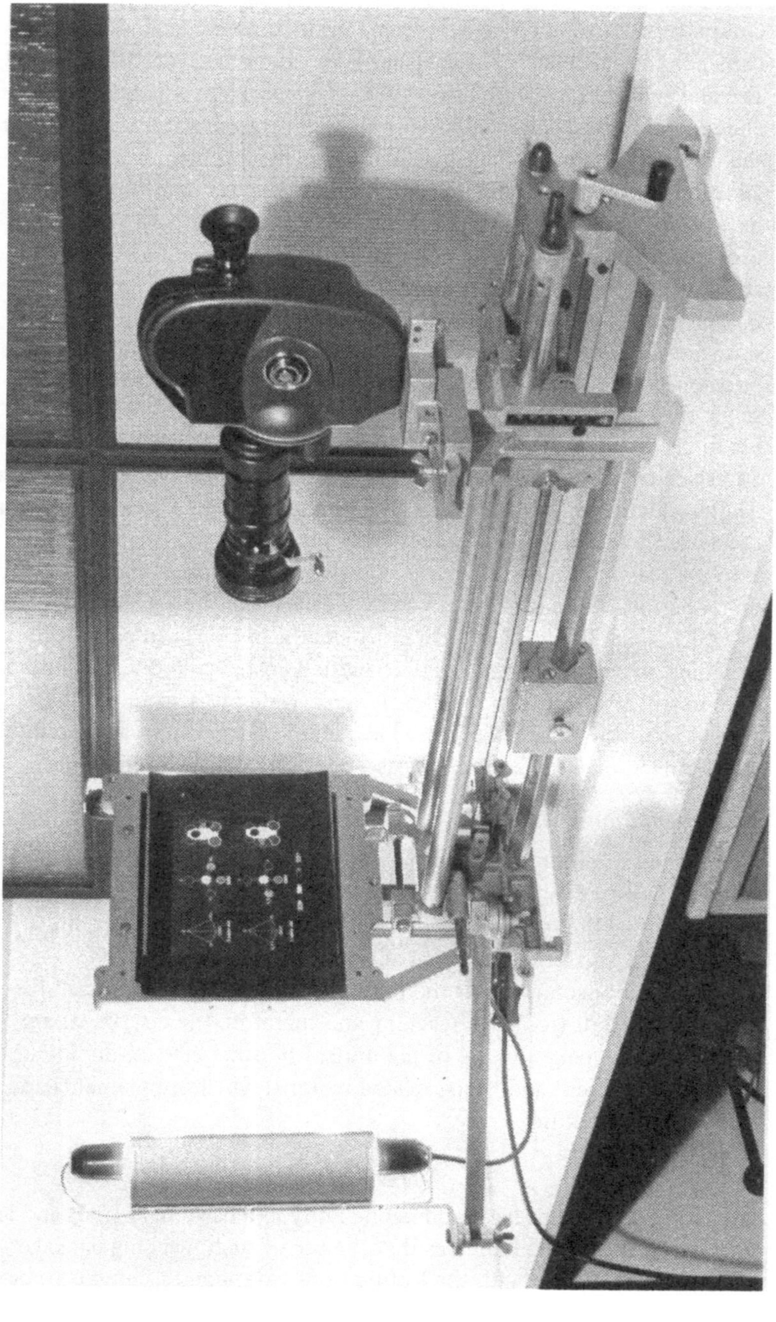

Figure 1.14: Camera Assembly for the Production of 16 mm Master Film from the Prepared Artwork

encouraged. The reaction from the students using PIP programs is testimony — their exam marks in this area were of a high standard, the significant differences occurring with the overall poorer student. This is illustrated in Figure 1.15 which shows the mark distribution plotted for a particular examination of 50 candidates in biochemistry. Included within this figure is the mark distribution for one specific question from the same examination paper, the one dealing with amino acids and protein structure.

Figure 1.15: Mark Distribution for a Biochemistry Examination Diet*

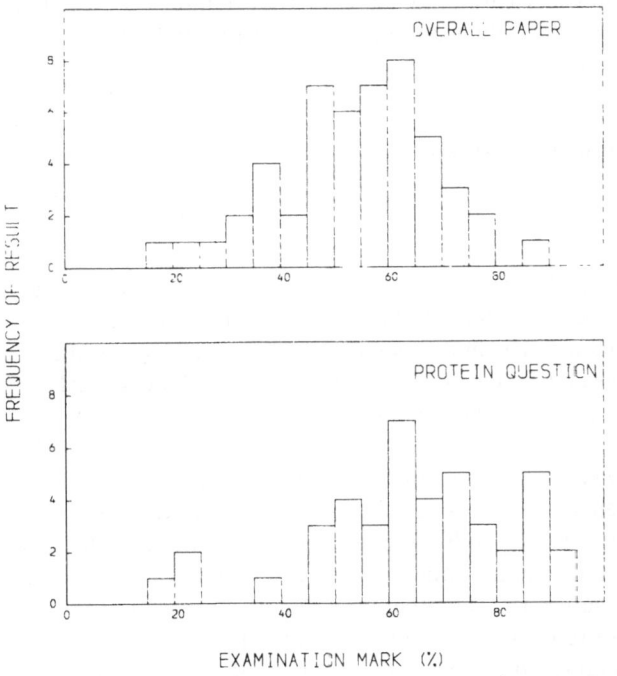

EXAMINATION MARK (%)

*Top frame represents the overall mark distribution, bottom frame represents the mark distribution for one specific question.

The format of the examination was essay questions/short notes with students attempting five questions from a total of eight. The overall median and mean were 55.5 per cent and 53.5 per cent respectively whilst the corresponding figures for the question on amino acids and protein structure were 65.0 per cent and 67.1 per cent from 42 attempts. From such an analysis it is evident that the learning package is effective in the teaching of the learning objectives to the students. One criticism of this type of analysis might be that this particular question in that examination paper was of a slightly easier standard than the others and that this would account for the observed differences in the mark distribution. Whilst it is extremely difficult to argue one way or the other on this issue, the indications are that the observed differences are real. This view is put forward on the basis of a comparison of student performance in an examination in which some students had used the learning package whilst others had taken a formal lecture course based on an identical syllabus. The marks distribution for this question for those students taking the lecture course was very similar to the overall distribution.

Topic 4: Porphyrin Biosynthesis

Design Phase

Target Population. The target population for this learning package was third-year BSc Science students and third-year Higher National Diploma students with an educational background of two years of a biological science course in which the fundamentals of biochemical metabolism are covered in the second year.

Aims and Relevance. As before, one of the aims of this package was to dispel any apprehension that the student may have as a result of the structural complexity of porphyrin molecules. The solution in this case was to generate the complex structure slowly in a step-by-step process as outlined in Figure 1.16. The other main aim of the package was to allow the student to learn each of the individual steps/stages of porphyrin biosynthesis *in vitro* and to be able to discuss structure–function relationships for this group of compounds.

Terminal Objectives. By the end of the learning package each student should, without reference to his or her notes or textbook, be able to:

(1) describe the general properties of porphyrins as a class of compounds;
(2) discuss the important structural elements of porphin;
(3) construct correctly a line drawing of porphin including in it the arrangement of double bonds;

Figure 1.16: Step-by-step Generation of the Structural Formula of a Porphyrin (Protoporphyrin IX)

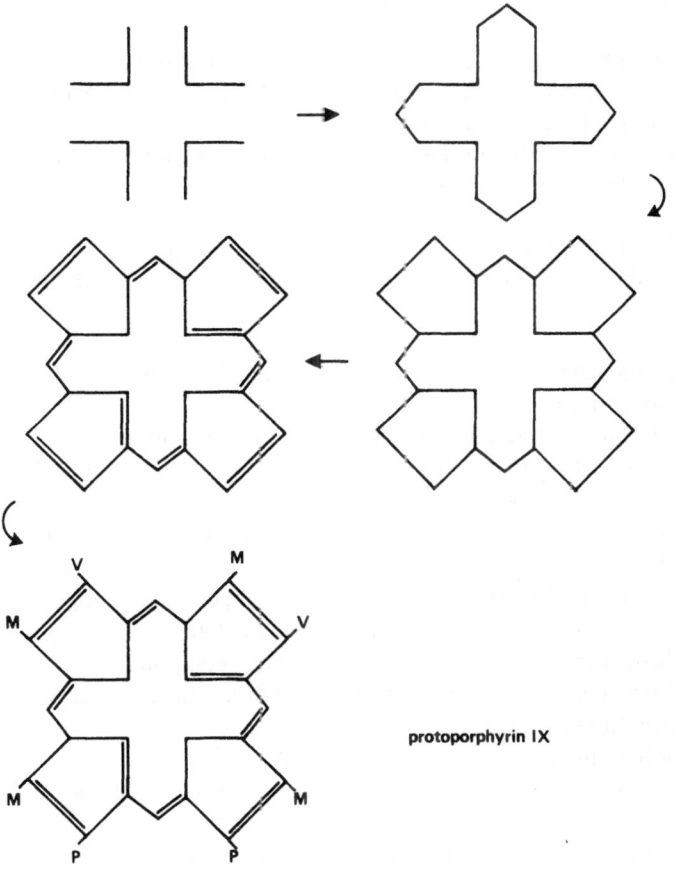

protoporphyrin IX

(4) state correctly the two starting materials of porphyrin biosynthesis;

(5) describe in detail the individual steps in the porphyrin biosynthetic pathway including the names of the enzymes involved and the co-factors, if any;

(6) define the term *committed step* as applied to metabolic pathways;

(7) indicate, by an appropriate diagram, the structures of uroporphyrinogen III and protoporphyrin IX and include in these the system of double bonds and the nature and spatial arrangement of the side chains;

    (8) describe the reaction in which iron is incorporated into protoporphyrin IX to give haem;

    (9) list at least four examples of enzymes which use haem as a co-factor;

  (10) summarise the structural characteristics of haem in haemoglobin and the cytochromes and comment on ligand binding to the co-factor;

  (11) distinguish between the haem group in haemoglobin and the porphyrin ring structure in chlorophyll.

Entry Behaviour. Before progressing onto this learning package, each student should, without reference to his or her notes or textbook, be able to:

    (1) define what is meant by the terms metabolic pathway and biosynthesis;

    (2) describe, in detail, the individual steps in the Embden–Meyerhof–Parnas pathway, the tricarboxylic-acid cycle and the mitochondrial respiratory chain;

    (3) define what is meant by the term co-factor (Bryce, 1979b);

    (4) discuss briefly the importance of co-factors in living systems;

    (5) outline the functions of the proteins myoglobin, haemoglobin, cytochromes, chlorophyll, albumin and catalase;

    (6) distinguish between cyclic and non-cyclic compounds;

    (7) distinguish between double and single bonds in chemical structures;

    (8) define the terms decarboxylation and oxidation;

    (9) identify correctly the porphyrin ring structure from other classes of compound;

  (10) outline the system used for enzyme classification;

  (11) describe the general protocol for isotopic labelling studies.

Learning Hierarchy. The learning hierarchy for this package is presented in the flow-chart format in Figure 1.17. Analysis of the type of learning required for each objective has shown this to be 'knowledge' according to Bloom's taxonomy.

Sequencing Sub-topics. The subject matter contained within this package is, by its very nature, linearly structured and this leaves very little freedom in the way in which sub-topics are sequenced.

Figure 1.17: Learning Hierarchy for the Learning Package on Porphyrin Biosynthesis

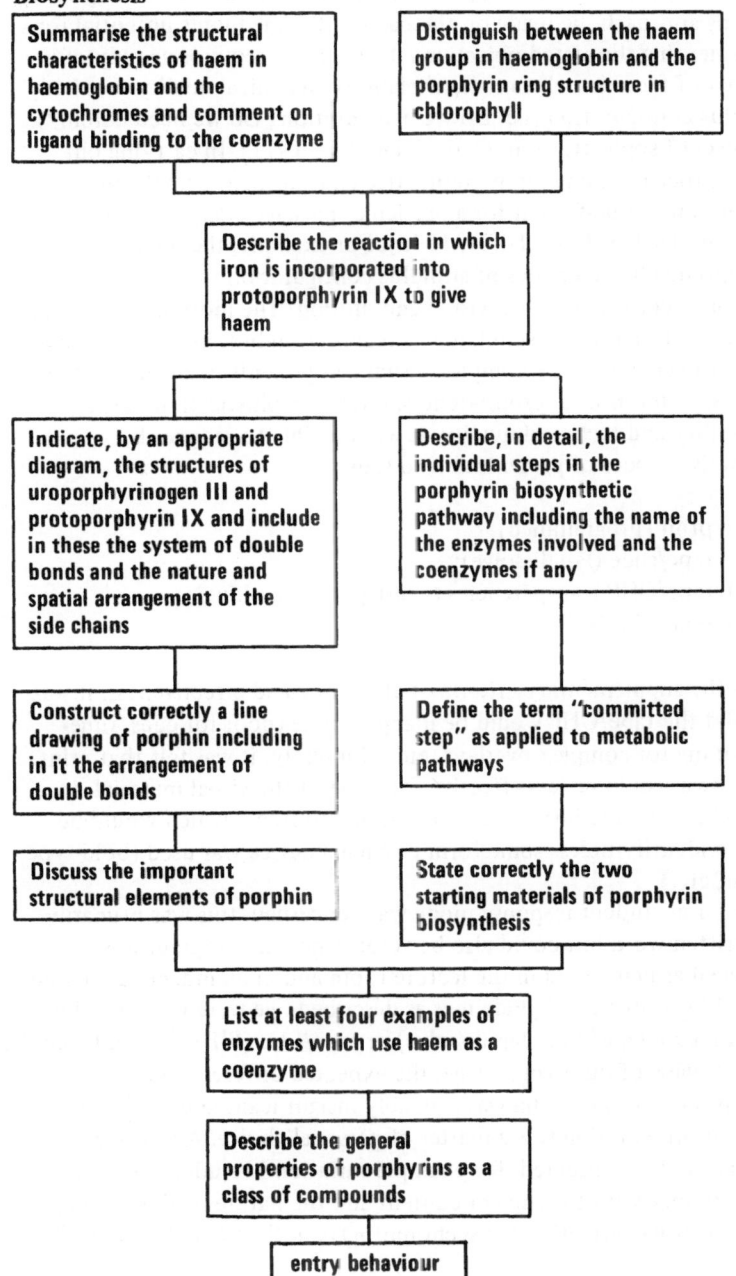

Summarise the structural characteristics of haem in haemoglobin and the cytochromes and comment on ligand binding to the coenzyme

Distinguish between the haem group in haemoglobin and the porphyrin ring structure in chlorophyll

Describe the reaction in which iron is incorporated into protoporphyrin IX to give haem

Indicate, by an appropriate diagram, the structures of uroporphyrinogen III and protoporphyrin IX and include in these the system of double bonds and the nature and spatial arrangement of the side chains

Describe, in detail, the individual steps in the porphyrin biosynthetic pathway including the name of the enzymes involved and the coenzymes if any

Construct correctly a line drawing of porphin including in it the arrangement of double bonds

Define the term "committed step" as applied to metabolic pathways

Discuss the important structural elements of porphin

State correctly the two starting materials of porphyrin biosynthesis

List at least four examples of enzymes which use haem as a coenzyme

Describe the general properties of porphyrins as a class of compounds

entry behaviour

Media Selection. In considering the most appropriate medium or mixed media for a specific teaching package, selection should, according to the systems approach, be made on the basis of the medium's potential for implementing the stated objectives. This in itself requires a study of a number of factors such as cost, availability, technical quality required, stimulus configuration (i.e. Do we need motion? Do we need colour?), and level of sophistication (Gerlach and Ely, 1971). In carrying out such a procedure, one rarely in practice ends up with a single most appropriate media-format for a particular package (Carpenter, 1971; Levie and Dickie, 1973; Hiedt, 1975). For example, the most appropriate characteristics of stimulus configuration for the present learning package are verbal, visual and motion. The motion, however, is not meant to represent simulated or real motion but, instead is used as a means of generating the complex visual components in a step-by-step process. In terms of appropriateness, level of sophistication, cost, availability and technical quality, it was possible to reduce the alternative modes of presentation to four:

(1) print (programmed);
(2) tape/slide (lap-dissolve);
(3) tape/OHP transparencies (overlaps and masking);
(4) tape/film (PIP system).

Using the supplementary criteria of efficiency and effectiveness, it was felt that the tape/OHP would be inappropriate (poor for cues, time-consuming for complex overlays, etc.). Similarly, it was felt that print on its own would be less effective for some of the visual material since small changes were introduced from frame to frame which would be slow to identify unless some form of cueing device was used (bold-type, colour, etc.).

As far as student response-mode was concerned, this was primarily the written-word but could also be verbal since the program was envisaged as being used in the lecture-room and in an independent study area. Using criteria and guidelines as described, we were left with the mixed-media tape/slide (lap-dissolve) or tape/film (PIP system). From the point of view of our own college, the expected difference in cost-effectiveness of the media was probably insignificant and so the final mode of presentation was a matter of personal choice. Additional criteria can be considered, for example variation in student ability, compatibility with the subject content and the nature of the learning task (cognitive, affective or psychomotor) (Cheek, 1977; Bryce, 1979d;

Mason, 1979; Herskovitz, 1979; Winn, 1979; Wager, 1980). Since it was intended that the program be initially used in the lecture-room, a 35 mm slide format was chosen as this provided better projection properties for group viewing than the 8 mm film of the PIP system. It was felt that if the lap-dissolve system was effective then this could easily be converted to PIP format and the latter be used in the independent study area on an individual basis.

In the context of media selection, it is worthwhile to note that there is evidence to suggest that pupil performance from different media can be associated with the attitude of the teacher to those media and this has become known as the Hawthorne effect (Tobias, 1969; Dodge *et al.*, 1974; Parsons, 1974; Hunt, 1977).

Student Evaluation. For this particular learning package, three different objective tests were constructed: (1) an entry test which covered the material implicit in the entry requirements and for which the student was required to achieve a standard of at least 90 per cent (ten test items); (2) a pre/post-test which each student took before and after covering the material content of the package; and (3) a program test which related to specific questions raised during the running of the program. The pre/post-test results taken together would provide an index of the learning gain, a parameter which was extremely useful in validating the actual package.

Production and Implementation Phases

The artwork for this package, in the form of line drawings and typed text, was converted to 35 mm lith slides. The image was coloured using a variety of self-adhesive transparent colour film (AGA filmolux: twelve colours) applied directly to the lith film. Additional shades can be obtained by combinations of two or more of the different colours. Using this colour film is an easy, albeit occasionally time-consuming, method of generating high quality multi-coloured lith slides. For accurate registration of the visual material during the dissolve sequences in the program there was an absolute requirement for slide mounts with registration pins.

Validation Phase

In the three years which this program has been implemented the results are very encouraging as typified by high increase (70–90 per cent) in the learning gain. Student reaction to this particular media-format is very good partly because the effects can be very striking and very professional-looking (especially if a dissolve colour change occurs) and partly because it is a little different (Hawthorne effect again). Other workers have recently found similar outcomes using this media in the teaching of undergraduate chemistry (Harpp and Snyder, 1977).

From a practical point of view the use of lap-dissolve can be very demanding. If, for example, as in this case there is extremely delicate registration involved (a line of text being superimposed by the same line in a different colour or modified in some other minor way) then it can be very time-consuming to set up the lap-dissolve projector system. One solution to such a constraint, apart from changing the program, is to copy the lap-dissolve sequence directly onto another more suitable format like film or television. This we have done with some success in the case of the program on porphyrin biosynthesis. The method employed was to project the lap-dissolve slides onto a rear-projection screen and film from the other side in a darkened room using 16 mm film. The major technical problem which we encountered in this process was one of exposure, and to achieve the desired exposure we were required to use high-speed colour film (Video News film, HS 7250 with a rating of ASA 400). This produced a 16 mm master from which 8 mm copies were produced by reduction printing and these copies were used in cartridge form on the Philips PIP system.

Concluding Remarks

In a recent article on the job satisfaction of university teachers, Gruneberg and Startup (1978) considered a number of aspects of the lecturer's job and they constructed a questionnaire to assess which of these yielded the greatest job satisfaction. High on the list was interest shown by students, and their capabilities as a lecturer and as a tutor. If this is the case then the future of the use of resource-based education is very bright. In Muir (1978), Irwin Edman (1896–1954) is quoted as saying 'Education is the process of casting false pearls before real swine.' Hopefully, however, at this stage in the book readers will accept the advantage of using learning objectives to sift out these *false pearls* and that employing a systems approach may make the student a little less porcine. The remaining chapters concentrate on a number of smaller areas of current interest and provide detailed accounts of the fields in general and of the authors' specific contributions to the subject.

Acknowledgements

I would like to thank Mr A. M. Stewart and Miss M. Gunn for their very keen interest in the design, production and validation of each of the packages described and for their inestimable help and guidance in this field. I would also like to thank all my colleagues who have helped in the implementation and validation phases and in particular to Dr David Button (Dundee College of Technology) and Dr Roger Griffiths (St Andrews University) for many helpful hours of discussion.

I would like to acknowledge the financial support from (1) the Research Committee of the college over the last four years, (2) the Nuffield Foundation for an award under the Small Grants Scheme for Undergraduate Teaching for development work on the PIP system and (3) the Scottish Education Department for a research grant for the computer-assisted learning aspects of the study.

2 SYLLABI CONSTRUCTION, CURRICULUM CONTENT AND RELEVANCE IN MEDICAL BIOCHEMISTRY

Krishnamurti Dakshinamurti

Introduction

Among the basic sciences included in the education of the medical student, biochemistry occupies a unique position. Everybody recognises that biochemistry is in the forefront of man's quest for new knowledge, not just esoteric but that which is vital in our continuing battle against diseases and in developing newer modes of treatment including enzyme replacement therapy and genetic engineering. It is in the nature of this discipline that no matter which clinical problem is investigated, one has to use the concepts and techniques of biochemistry to understand the disease at the molecular level. Advances in biochemical science are the root of progress in medicine (Kornberg, 1978). The impact of biochemistry on biology and medicine has been enormous. In view of this brilliant record, it is surprising that the role of biochemistry in the medical curriculum is questioned.

A superficial consideration might indicate that there are separate entities, one dealing with progress in the medical sciences and the other with the biochemical content of the medical curriculum. The latter is usually linked to another question 'What biochemistry should a practising physician know?' The definition of the term *practising physician* then would determine what biochemistry one uses. There is a wide spectrum of practising physicians from the front-line general practitioner to the specialist, and depending on whom one chooses as the yardstick the answers would vary. Also related to this is the general philosophy of medical education. The Flexner model was directed toward the training of the physician-scientist. In times when this was fashionable the aim was to train the physician in such a way that he could, if he chose, adapt to a teaching/research-oriented clinical career. Most agree that such training would result in a better clinician. In the latter half of the 1960s new pressures began to be felt on medical education. The social unrest of the times fashioned the new thinking, particularly that of the students. This was compounded by the consumeristic attitude of the public toward medical care. Health care delivery became the primary concern. Anything that could cut the cost and time in training new physicians was fair game for both the politicians and administrators. The dearth of physicians in rural areas or small towns was confused with a general shortage. Great value was placed on shortening the time spent in training. This led to the demand for *relevance* in the medical curriculum to the practice of medicine in *general practice*. It is surprising that the problem of shortage of physicians was not addressed by significantly increasing the enrolment in medical schools. These considerations lead us to certain crucial decisions. What kind of a physician are we going to train? Should we train different kinds of physicians with varying levels of academic training? Is the prime need the provision of more 'bare foot doctors'? In as much as the consensus is the provision of a single category of primary training we have to examine the curriculum that would best educate the prospective physician.

As professional biochemists we must have the academic integrity to assert what we consider to be essential to the understanding of the living process (Murray, 1979). What role should biochemistry play in the medical curriculum? In answering this the complex nature of the discipline must be kept in mind. Most of us who refer to *biochemistry* have various aspects of it, like biological chemistry, clinical chemistry or chemical pathology, in mind. Areas like molecular biology, molecular genetics, bioenergetics, biomembranes and metabolic controls are at the forefront of the current thrust in biochemistry. Research expands and pushes forward the horizon of fundamental knowledge in these areas. Even when such research is not disease-oriented, the overflow effect of the advances gained on the development of curative and preventive medicine is obvious enough. The medical student needs to be exposed to this aspect of biochemistry. Clinical chemistry deals with the determination of various constituents in body fluids and tissues, the development of such methodology including the delineation of *normal* values and the identification of problems in interpreting the analytical values. Another aspect of clinical chemistry which some might refer to as 'clinical biochemistry' or 'chemical pathology' deals with the chemical basis of aetiology of disease processes. We recognise that not all diseases can be explained at the molecular level. Needless to say, that the study of the correlation of biochemical abnormalities with the clinical symptoms and their pathological consequences, to the extent we understand various diseases, is essential in the training of the medical student.

Biochemistry is only one of the several subjects in the medical curriculum. Other disciplines — basic or clinical — are equally important in the training of the prospective physician. One has to keep this in perspective to apportion properly the available time to the various disciplines that form the medical curriculum. Also, the profusion of published literature is not limited to biochemistry. Thus we have to be realistic in determining the time to be allotted for the study of biochemistry as well as in selecting what 'biochemistry' needs to be included in the curriculum.

In Canada, an increasing proportion of entering medical students in most medical schools have already taken a biochemistry course. Medical schools like ours have included an undergraduate course in biochemistry among the academic prerequisites for admission to first-year medicine. This is a very welcome development. It helps to reduce the stress of learning during the first year of medical education which is overcrowded. A few words about the 'prerequisite biochemistry' will not be out of place here. This course should be a 'general biochemistry' course taught by the Faculty of Science and intended for students who would continue in science or branch off into other fields like medicine, agriculture, etc. This course should have laboratory work as a component. The advantage of this approach is that the student would be exposed to all areas of biochemistry without an inherent bias towards 'human' biochemistry. If students have already taken a course in pre-medical biochemistry, do they need another course in bio-chemistry in the medical school? With the answer obviously in the affirmative, we have to address ourselves to the next question, what 'biochemistry' should we offer to the student of medicine?

An attitude which suggests that students should receive the type of course that they and the community needs (Schwartz, 1980) betrays much confusion. This is as irrelevant as proposing an advanced course in biochemistry which has very little to do with the biochemistry of health and disease. What is needed is a course which would prepare medical students for future years of medical education. It is also recognised that in later years of practice the clinician will have to draw on concepts and problem–analysis skills learned in basic science courses to keep up with the advances in medicine. There have been many attempts at developing curricula to make biochemistry relevant to the study of medicine (Saffran, 1971; Rubinstein, 1972). A series of case presentations forming an integral part of the lecture series has been used in these. This format has been designed to motivate the students. However, the various selection processes used in the admission of students to medical school are expected to produce a body of properly-motivated students.

Our course in biochemistry is not a course either in clinical medicine or in chemical pathology. It would not be very productive to discuss this at this early stage in medical education, although there are dissenters (Blanchaer, 1975) of this view.

Curriculum Design

In the succeeding sections I describe a syllabus in biochemistry for first-year medical students developed by the Biochemistry Department of the University of Manitoba. An outline of this course has been published earlier (Dakshinamurti, 1979). It was developed as a course on 'human biochemistry', primarily oriented toward providing an understanding of the normal biochemical process in the human body in which the functions of the various organs and tissues are integrated. The comprehension of the principles of metabolic integration would contribute to the students' understanding of the biochemical basis of various disease processes. It was decided not to teach biochemistry in a strictly case-oriented fashion as this would appear contrived. Certain areas of biochemistry are of such fundamental importance that an obvious clinical relevance is not necessary. However, we have introduced interesting clinical examples where we deemed it appropriate. Even rare diseases were used as examples to illustrate the use of biochemistry as a scientific investigatory technique.

A realistic assessment of the biochemical background of entering students indicated that they had adequate knowledge of biochemical language and the elementary 'road maps' of metabolic interconversions. The areas of biochemistry to be included in this course were determined after much analysis, reflection and discussion by the teaching faculty. A large number of clinical case histories were individually analysed for the biochemical topics associated with each. To this list of topics was added 'basic' biochemistry of intrinsic merit without obvious clinical relevance. The detailed outline of the course was extensively discussed by the Teaching Committee before its adoption. In view of our distinctive approach in developing this course we felt that we could not select any of the available biochemistry textbooks as a recommended text for our students. We have provided printed lecture notes to the students. Individual instructors of this course prepared the lecture outlines for each section which were reviewed by a group of three consisting of the instructor charged with its preparation, another instructor knowledge-able in the general area and a third member whose area of expertise was different. This was designed to provide a proper balance and avoid too much specialisation. Following this, the detailed lecture notes were prepared and the same scrutiny and review were applied for these as well.

Once the course content was decided upon, the next step was to allocate a reasonable distribution of the assigned lecture hours. We decided that the best way of presenting biochemistry would be through a combination of didactic lectures, quarter class tutorials (T_4 with about 25 students per group) and small-group problem-solving encounters (T_8 with about 12 students per group). We also use the biochemistry teaching time to initiate students in a 'Literature Search and Evaluation Project (LSEP)' (see later). There are 109 student-contact hours of biochemistry instruction in the first-year medical curriculum. These are distributed as follows: lectures, 40 hours; T_4, 14 hours; T_8, 24 hours; LSEP, 6 hours; and biochemistry problem-solving time, 25 hours. The list of topics in biochemistry and the time allotted for each section is given in Table 2.1. Each section starts with one to three hours of lecture followed in some instances by a T_4 (one hour) and in all instances by a T_8 (one hour). Attendance at lectures as well as at tutorials by the student is optional.

Course Overview

Sufficient basic biochemistry as is necessary to understand the integrated functions of the human body and its normal metabolic regulation is provided. As a corollary, the site of impaired regulation in diverse metabolic pathways becomes apparent as the biochemical lesions of the disease process. A knowledge of the normal metabolic process provides the rationale for clinical tests that are employed diagnostically.

The methods used in studying metabolism in the intact human organism are introduced, and this is followed by other investigatory tools on whole organs, isolated tissues or cells, progressing to the use of pure enzymes in understanding the biochemical manifestations. Inter-organ co-ordination in preservation of homeostasis is illustrated in the following section dealing with acid-base and electrolyte balance. The sections on proteins, enzymes and membranes deal with the molecular basis of essential processes such as oxygen, nutrient or metabolite transport and the regulation of enzyme activity. The molecular basis of intestinal and renal absorption defects are included.

'Energy metabolism' deals with concepts like energy balance, basal

Table 2.1: Topics Presented in Biochemistry

Topic	Lecture	T_4	T_8	*P.S.L.
1. Introduction to Metabolism, Methods for studying biochemical abnormalities	3	1	1	
2. Body fluids, electrolytes, acid-base balance	3	1	1	
3. Proteins	2	1	1	
4. Enzymes including clinical enzymology	3	1	1	
5. Membranes	2	–	1	
6. Energy metabolism	2	–	1	
7. Molecular mechanisms of movement	3	1	1	
8. Molecular basis of inheritance	2	–	2	
9. Nucleotide metabolism	1	–	1	
10. Digestion and absorption	2	–	1	
11. Mode of action of hormones	1	–	–	
12. Carbohydrate metabolism	3	–	3	
13. Lipid metabolism	3	–	3	
14. Protein metabolism	2	–	1	
15. Amino acid metabolism	2	–	1	
16. Vitamins	3	–	2	
17. Mineral metabolism	2	–	2	
18. Metabolic Integration	–	4	–	
19. Review	–	4	–	
20. LSEP	1	–	1	4
21. Problem solving	–	–	–	28
	40	**13**	**24**	**32**
Total = 109 hours				

* Problem Solving Library

Source: Dakshinamurti, K. (1979) *Biochemical Education*, 7, 70.

metabolic rate and the regulation of energy expenditure and appetite. The biochemistry of skeletal and connective tissue is presented in relation to their function as components in the system involved in movement. 'Molecular basis of inheritance' bridges the concepts of molecular biology and classical genetics and provides an understanding of how the genetic information in DNA through its expression in proteins relates to inborn errors of metabolism and polygenic diseases as well as biochemical individuality. The section on nucleotide metabolism illustrates the effects of derangement of the normal feedback regulation in the cell.

'Digestion and absorption' deals with the alimentary tract as the internal surface of the body and its role in absorption and excretion. The sections on carbohydrate and lipid metabolism emphasise the integration of these metabolic pathways and their control by metabolites, hormones and the nervous system. This provides the basis for understanding various disorders like diabetes mellitus and obesity as well as the various hyper-and hypo-lipidemias. The functions of essential fatty acids and prostaglandins in metabolism, the role of complex lipids in membrane function and lysosomal disorders are also considered in the section on lipid metabolism.

'Protein metabolism' deals with concepts such as nitrogen balance, essential amino acids and the biological value of proteins. The pathways of amino-acid metabolism and the utilisation of amino acids for production of specialised products or for specific purposes are also included. 'Vitamins' introduces concepts like vitamin requirement and genetic abnormalities associated with vitamin function including vitamin dependency and vitamin toxicities. The metabolic function of various vitamins and the clinical features of vitamin deficiencies are also considered. 'Minerals' deals with concepts like mineral balance and factors controlling mineralisation of bone. The last section of the course deals with clinical problem-solving based on the biochemical information provided in a series of four tutorials. This reiterates the metabolic inter-relationship as between different pathways and between different organ systems.

The T_4s are used in a structured or unstructured way as determined by the instructor. The main effort here is to go over concepts requiring reiteration or clarification of student's residual difficulties. No new material is introduced in these. The objectives of the T_8 are: (1) to illustrate the application of biochemical information and concepts to a wide variety of problem-solving situations related to medicine, (2) to encourage students to develop problem-solving skills both individually and as part of a group enterprise and (3) to reinforce and broaden the basic material. Different formats, including *simulated clinical problems* (Blanchaer, 1975), all based on case histories with primary data have been used. Information in the form of a case history with required data (clinical, biochemical, etc.) are provided to students in written form well in advance of the T_8 for study and preparation. While four groups comprising half the class are engaged in the T_8 sessions the other half of the class has free time intended to be used for the preparation of biochemistry tutorials.

The students are aware of the emphasis placed by the faculty on small-group discussion as a learning method. All entering medical students go through an orientation program during their first week in medical school when they are introduced to small-group learning encounters. During this time they acquire the skills necessary to participate effectively in such classes. Most members of the Biochemistry Department have attended programs on small-group teaching organised by the Faculty of Education of this university (see also Chapter 7). The list of topics offered in the T_8 is given in Table 2.2.

These tutorials offer the opportunity for active discussion and decision-making by the participants. In the T_8s an attempt is made to integrate into the tutorial material biochemical topics covered earlier in the course. In addition to this, we have tutorial sessions (T_4s) on various aspects of metabolic integration. Topics like effects of starvation, metabolic effects of alcohol, diabetes and cardiac arrest offer excellent opportunities to discuss metabolic interrelationships, their integration and control.

Objectives of the LSEP

The objectives of the LSEP are: (1) to learn procedures of effective literature retrieval, (2) to develop the ability to evaluate a research report in the form of articles published in a scientific journal and (3) to learn to judge what is essential for the preparation of a report on the literature search. The topics are sufficiently specific that students do not have to examine all the current literature in a whole field. The instructors also ascertain that appropriate articles are in fact available in our library holdings before assigning a topic. When the LSEP is introduced to the students they are instructed by the medical librarian in a highly-effective method of literature search (Kerr, 1970). Each group of about twelve students is assigned a topic which they search in

Table 2.2: Topics Discussed in T_8 s

1. Protein losing enteropathy – the use of radioisotopic tracers administered through different routes in investigation of a metabolic abnormality.

2. "Willie" – the consequences of acute glomerular nephritis on acid-base and fluid balance.

3. Sickle cell anemia – the molecular pathology and basis of treatments.

4. Serum enzymes in various diseases – diagnostic use of serum enzyme assays.

5. Watery stools – electrolyte changes as a result of cholera toxin exposure and the concept that a simple molecule like G_{M1} ganglioside can act as a membrane receptor.

6. Luft's Syndrome – hypermetabolism due to uncoupling of muscle mitochondrial oxidative phosphorylation.

7. Connective tissue diseases – the relationship between connective tissue diseases and the pathway of collagen and glycosaminoglycan metabolism.

8. Lesch Nyhan Syndrome – "Twins with hyperuricemia", including prenatal diagnosis.

9. Genetic engineering – the facts and fiction of genetic engineering.

10. Orotic aciduria – genetic defect in hereditary orotic aciduria.

11. Enterokinase definciency – the role of enzyme activation in digestion.

12. "Dulcina" – the significance of reducing substances in urine, the techniques used in testing for them and the interpretation of the tests.

Table 2.2 – *continued*

13. "A test of tests" – human problems associated with hyperglycemia and the choice of techniques used in testing for it.

14. "Where, what is it?" – glycogen storage diseases.

15. "Linda" – carnitine deficiency and fatty acid oxidation.

16. Hyperlipoproteinemia – endogenous hypertriglyceridemia; hyperchylomicronemia.

17. Primary hypercholesterolemia – LDL receptor hypothesis.

18. "Diabetes" – laboratory findings in this disease and the underlying pathways of carbohydrate and lipid metabolism.

19. Protein metabolism – nitrogen balance.

20. "Elsie" – a case of homocystinuria caused by pyridoxine dependency.

21. Megaloblastic anemia, interrelationship between folic acid and vitamin B_{12}; polyglutamate conjugase.

22. β-Hydroxy isovaleric aciduria, β-methylcrotonyl glycinuria – problems of branched chain amino acid metabolism.

23. Mineral metabolism – analysis of flow routes of calcium and phosphate; techniques of determinations and interpretation of data.

24. Lead poisoning – absorption, excretion and biochemical effects of an environmental toxin.

Source: Dakshinamurti, K. (1979) *Biochemical Education*, 7, 70

the library. The time required for library work is also provided within the biochemistry hours. Each group presents and discusses its findings at an assigned time. The instructor who provided the topic is present at this discussion as a resource person. Each group decides on how to use its manpower in the various tasks associated with collection, evaluation and presentation of data.

Concluding Remarks

During the two years of experience with this course we have evaluated our students' understanding of the principles of human biochemistry and their ability in problem-solving. A positive effort was made to avoid the educational trap of making the course extensively factual. We have been successful in achieving the goals we set ourselves at the start of this program. With experience comes the confidence that we have in our course the right distribution between basic and clinically-relevant biochemistry, and between didactic and small-group teaching with the emphasis primarily on problem-solving. The syllabus we have developed could be modified to suit individual circumstances of any medical school as it applies to pre-medical student preparation, as well as to the availability of staff manpower.

I would like to close this discussion with some personal reflections on biochemistry in medical education. We have come a long way from the traditional lecture *cum* laboratory routine in general biochemistry of presenting various aspects of human biochemistry to first-year medical students. There is recognition that the students' contact with biochemistry should not stop with this course. There are tremendous opportunities as well as challenges for a biochemistry faculty with proper motivation for integrating biochemistry both horizontally and vertically with other subjects of the medical curriculum during the entire period of medical education. We have in our faculty the beginnings of such successful integration with subjects like neuroscience and gastroenterology. A notable achievement has been the institution of an intensive 16-hour program in clinical biochemistry during the third year. This provides an excellent opportunity for the student to establish correlations between actual laboratory biochemical data, clinical presentations and the pathology of selected cases.

3 A CASE-ORIENTED APPROACH: DESIGN, IMPLEMENTATION AND RESULTS

Rex Montgomery

Most students who take courses in biochemistry do not intend to become biochemists. Their purpose is either to prepare for other areas of study for which biochemistry serves as a base of understanding or else, in the rarer cases, the biochemistry course adds to the cultural balance of their education. For example, biochemistry might be the source of answers to such questions as: What is recombinant DNA all about and why should society be concerned? What damage may be done to normal cells by radiation and should I join those opposing the building of nuclear-power plants? Is cancer being significantly increased in an aetiological sense by environmental pollutants? These are the real questions of an educated public who is searching for answers. Balanced argument is provided in part by understanding the molecular aspects of the problems. The questions are different, but the purpose is the same, for those students in the biological or health sciences. In all cases the inquiring mind is searching for answers to questions that are more extensive than the molecular basis of the living system but require knowledge of the biochemical principles involved. Such students are not as impressed by the logic of a biochemical principle or the aesthetic beauty of a crystal form as they are by the light shed upon a problem in an area of their concern through the application of biochemistry.

It is not appropriate to collect all these students for whom bio-chemistry is a minor subject into one group and the intended practitioners of biochemistry into another, applying two types of education to them, just as Plato identified two classes of citizens in his city-state. One class of his citizens received the doctrine as a body of knowledge that was not to be questioned, in contrast to the policy-makers who were taught the art of continuing inquiry. The educational programs of all students should enable the development of procedures of scientific inquiry, applicable to any unresolved problem (Gagné, 1963). The approaches to teaching biochemistry with such an objective of terminal behaviour are numerous, probably as numerous as the number of biochemistry teachers. The steps in the learning process however are common to all the approaches: the student acquires a knowledge of the principles of biochemistry, without which the deductive thinking process cannot be expected; the understanding of how the principles are used is developed in parallel.

Each field of study or practice has a language and a basic set of rules, without which the 'game' cannot be played or discussed; as in the game of bridge there develop sub-sets of conventions whereby some group of players exchange information differently from other players of the same game, but all the players have the same objectives to be achieved within the same body of rules. Thus, the biochemistry student must be taught the principles and the language of the sub-sets of the field, such as enzyme kinetics, virology, physical biochemistry and so on, before a reasonable practice of scientific inquiry can be exercised. It may be expected, though not presumed, that the student will have received previously some general scientific instruction, has been exposed to some chemical, biological and physical principles and has seen how these are derived from an experimental approach through observation, precise descriptions of these observations and the generation of reasonable models.

In trying to develop biochemistry in the minds of any student as a subject for scientific inquiry, which is not achieved by the mere knowledge of the principles, the careful consideration of teaching strategy is called upon. It requires the interpretation of biochemical principles, the selection of the relevant facts from the total data and eventually bringing to the student the ability to use facets of both fact and principle in understanding the problem at hand. For example, a patient presents with a history of disease and many facts are generated in the physical examination and laboratory studies. How can the biochemistry of the problem aid in arriving at a diagnosis and treatment? What are the significant facts in the mass of data? Again, a cell in tissue culture responds in a particular way to the presence of a toxic substance. What is the experimental approach to the resolution of the mechanism of action of the toxic material? These uses of biochemical training bring the subject alive and usually stimulate the student to broader inquiry.

Design of a Biochemical Course Using a Case-oriented Approach: The Iowa Experience

For many years the biochemistry course for medical students at the University of Iowa called for a didactic course of lectures, which may have included periodic lectures of a clinical nature given by invited physicians, and a parallel laboratory course. The biochemical principles were illustrated by clinical examples and the laboratory exercises drew from clinical chemistry as much as possible. Many variations were played on this theme without any great success in stimulating the student to learn the subject as a useful part of the armamentarium of any future clinical practice. Following the surge of interest in learning objectives (Magar, 1962, 1968), all the courses in the medical curriculum were analysed and their behavioural objectives identified. The writing of these objectives probably benefited the faculty as much as the students, and many of the latter eventually looked upon the final document as a contract between them and the faculty. In the opinion of some faculty members, the objectives inhibited reading and study outside of the stated behavioural expectations. The documents, however, enabled an examination of the curriculum in reviews of clinical subjects across all courses, such as infectious diseases, gastroenterology, clinical chemistry and cancer. The results of these vertical reviews pointed clearly to omissions of important topics in some courses and the multiple treatment of some topics, such as blood clotting, in many courses, sometimes with contradictions in interpretation of both fact and theory. It was also apparent that the clinical years of the curriculum used little of the basic science material.

A similar conclusion was derived from a survey of the biochemical topics in cases presented in issues of *The Lancet* during 1970–3 (Wills, 1974). Approximately 38 per cent of the clinical presentations included mention of an important biochemical topic. Hormones, immunology, electrolytes/trace metals and nutrition were most frequently cited, while metabolism was a minor topic; about one-half of the metabolic topics related to porphyrins and bile acid. With the more recent increase in diagnosis of inborn errors of metabolism by the improved instrumentation of chromatography and mass spectrometry, this balance of biochemical topics in the medical literature may have changed towards more metabolism, but it still behoves the biochemistry teacher of professional students to contribute meaningfully to their ability to practice (Baggott, 1976).

During the 1960s the philosophy of teaching by scientific inquiry developed in the minds of some faculty members as did the cry of relevance from the student body. The one did not relate to the other except to call for some change in the method of teaching. Many faculties in medical schools across the world were caught up in this desire to change the teaching techniques. In several cases the changes were directed to provide alternate ways of teaching the students the principles of biochemistry and the facts of the biochemical language.

Games were played that made the mastery of facts and the understanding of the biochemical principles more palatable. Examples of these methods are found in several chapters of the present text. The essential problem for the non-biochemists is how to use this information in their professions and it was to this point that our attention was directed, taking medicine as the profession for study.

Problem-solving in clinical medicine (Culter, 1979) involves the collection of data on the presenting patient and selection of the data into groups that relate to each other in some possible way, thus leading to a tentative diagnosis or diagnoses that would account for the data. Such diagnoses may need confirmation and any unresolved observations will call for further study following the assessment of a treatment plan, by which time the original data base is much expanded. This general cyclical procedure of problem-solving is not dissimilar to that in many other fields, be it research or criminal investigation. For the purpose of teaching biochemistry from such a clinical data-base it is necessary to teach the student how to select the data of biochemical relevance and to recognise that the program of teaching is neither a mini-course in internal medicine nor a patient management problem. It is, however, necessary for reasons of brevity to use some medical terms, which can only be avoided if equivalent phrases are substituted each time they are used.

The design of the case-oriented biochemistry course was, in retrospect, a natural consequence of a review of the biochemical components of the medical curriculum and came at about the same time as Weed's patient-oriented medical records were being introduced (Weed, 1969). The general approach designed to teach biochemistry was to present the language of the area or principle and to use this information to analyse the biochemical components of a case, these analytical sessions replacing the usual time devoted to laboratory exercises. It was considered important to conduct such case analyses in the first week of the course, the earliest cases therefore being selected for the minimum knowledge base of the student. The vertical review of the curriculum showed that nutrition was diffused throughout the medical curriculum and was poorly represented as an identifiable learning objective, yet it was a subject very much in the minds and publications of the lay public and the entering student. A focused nutrition unit in the biochemistry course would present at least a foundation from which to build the specialty areas, such as paediatric nutrition, hyperalimentation and diseases of absorption. For these reasons, nutrition was studied first in our biochemistry course. Having developed this lead in the sequence of topics, the other areas for study followed the sequence that was comfortable for our style of teaching, with the possible exception of acid-base, fluid and electrolyte control. Because abnormalities of fluid and electrolyte regulation are so common in many diseases, this topic was introduced as early in the course as possible, which meant that the properties of proteins and the regulation of enzyme activity were necessarily precedent to it. Based on the sequence of these first four topics, a wide range of case discussions could be selected.

Case Analyses

First and foremost it must be remembered, and the student reminded, that one of the overall objectives of the course is to understand the principles of biochemistry. The second overall objective is to prepare students to use this knowledge. The third aim is to prepare the student to learn more about biochemistry by demonstrating in the time available that it forms one of the bases of future understanding of biological systems and disease. Speaking of the general topic of internal medicine, it has been stated that the advances of the future will rely upon the continued support of the basic sciences (Gray and Soffer, 1980). Fortunately over the last few decades our academic clinicians have increasingly taken up this banner for the basic sciences.

Cases were selected in which a biochemical principle was involved and a particular use of that principle exemplified. The disease was identified in the case description since we were trying to remove the impression that medical diagnosis was being taught. In several respects this differs from other uses of case histories (Winkler, 1978; Baggott and Trojak, 1978; Smith and Jepson, 1972). Our cases were chosen because they represented a biochemical data-base from which interpolations and extrapolations could be attempted. For example, in a case of galacto-saemia in a newborn girl of non-consanguinous healthy parents, the galactose 1-phosphate uridyl transferase in the erythrocytes was absent. Would this girl, when an adult, be able to secrete milk of a normal lactose content? What differences in the child's history would have been noted if the deficient enzyme had been galactokinase? The first question calls for the application of direct knowledge of the biochemical metabolism of carbohydrates to answer a reasonable question from the mother who is concerned for the future of her child and, although outside the analysis of the immediate case, clearly exercises the student in constructive use of his knowledge in a real-life situation. The second question is less directly applied to the case and asks the student to proceed in the same type of logical analysis of the case history except that a different inborn error is present. In another case, identified as a problem of dehydration, a patient, who had been rendered unconscious by a blow on the head, was fed a stated diet by gavage. The blood and urine data in the final days of life were given, the critical values being the daily urine volume and the elevated levels of BUN, Na^+ and Cl^- in the blood. The questions of nutrient and fluid needs for the comatose patient called for direct application of the principles of sound nutrition. The suggestion that death was due to dehydration and not damage done by the blow called for a more searching analysis of the fluid-need for proper kidney function, again attempting to extend scientific inquiry in a logical manner.

It is relatively simple to find cases of uncommon inborn errors of metabolism, which serve to demonstrate the effects of a single enzyme deletion or reduction in the many interrelated processes of the human body. Furthermore, it is sometimes quite revealing to see the extraordinary, and at first glance unrelated, consequences of such diseases. The cataracts produced in uncorrected galactosaemia is one example as is the increase in uric acid resulting from high-fructose diets or, in some cases, of fructose intolerance. Nevertheless, it was considered important to identify as many common diseases as possible for analysis, otherwise the relevance of biochemistry would be dimmed by the statistics of its usefulness. One comes to realise that if the common diseases were well-enough understood in biochemical terms they might soon become uncommon problems. Poliomyelitis and scurvy are two examples that quickly come to mind. These restrictions make the search for cases more difficult. However, the common diseases or conditions of diabetes mellitus, alcoholism, obesity, vitamin toxicity, jaundice, pancreatitis, gout and leukaemia are understood to a large extent at the molecular level and so serve widely in biochemical case-analysis.

In all of the cases a reasonably complete data-set is presented to give the student the opportunity to search for the biochemical consequences of the data, if any are known. Headaches, fevers and loss of appetite do not readily translate into biochemical parameters, but neither do some of the more common clinical-chemistry data because of the extraordinary ability of the body to compensate for changes from a narrow range of normal values.

Having selected an appropriate case, every twist of the biochemical content is developed, whereby the aberrant metabolism is compared to the normal condition and the total effects of the disease on the body are analysed. Unlike the *in vitro* experiment or the *in vivo* experiment with micro-organisms or laboratory animals, the indicators are limited. Body fluids can be analysed, as can biopsy samples of some tissues. After adding the data from the history and physical examination and diagnostic radiology, electrophysiology and nuclear medicine scans, the molecular status of the patient is unlikely to be further clarified. However, the data allows one to teach almost all the facts of mammalian biochemistry and illustrations of principle can be covered. The exceptions are those situations in which any dysfunction is incompatible with life. Such problems are then illustrated by cases of patient reaction to pharmacological compounds, toxins or poisons. Thus, oxidative phosphorylation is discussed under malignant hyperthermia in the rare individual who reacts in this fashion to the anaesthetic halothane, or in cases of uncoupling resulting from snake bite.

The organisation of the case analysis is dependent somewhat on the format of the required student response. Every aspect in the objectives of the course can be covered using questions that call for written responses from the student. The language and the principle(s) can be taught by appropriate questions before deductive reasoning is expected. Wherever possible the cumulative biochemical knowledge covered in the course is drawn into the questions so that the student develops a growing appreciation of the integrated nature of the biochemistry of the whole body. The final cases for study are comprehensive and the biochemistry is drawn from all areas in the questions; such cases are acute starvation, pancreatitis, diabetis mellitus, chronic alcoholism and haemachromatosis.

Small-group discussions of the cases have been most rewarding for the teachers in that the growing process of scientific inquiry becomes apparent. The pathways of logic of both students and teacher can be compared, each learning from the other. For example, the case history can be first analysed word-by-word for biochemical leads, the positive leads then being grouped, much in the fashion of clinical diagnosis, to examine one or another principle, the dysfunction of the normal, and the implication of the dysfunction in the whole body. Additionally, preparation for the discussion can include bibliographical research into the most recent published work and any additional related areas that may well be outside biochemistry. This brings to the fore the interdisciplinary nature of many sciences as well as the need for expanding our knowledge to other basic and clinical sciences.

Examinations are designed on the basis of case analyses. The questions attempt to be objective, searching for an evaluation of facts and principles of normal human biochemistry, but in addition they test the ability of the student to abstract the biochemical components applicable to the case in question. Evaluation of the student is arrived at in part from the small-group discussions during the course and in written responses on examination that require short answers, calculations or selection from multiple choices. We have relied to a significant extent upon the multiple-choice format for several practical reasons, two of which are the nature of the Medical National Board examinations in the United States of America, in which the students need practice, and in the application of computer scoring that permits the grade to be derived in the short time allowed by the university calendar. In spite of the inherent limited flexibility of the multiple-choice questions, which translates for the faculty member into the difficulty of designing a good examination, it is possible to search objectively the student's knowledge and ability to identify the biochemical implications in the data of a case history. Management of objective test item banks for biochemistry has been described (Bryce, 1979a), as have self-testing computer programs (Morgan, 1979a), open book problem-based examinations (Vella, 1979a) and exercises of a more quantitative nature (Montgomery and Swenson, 1976).

This method of presenting medical biochemistry has been used for eight years at the University of Iowa. The passing grade is set at 70 per cent, averaged over all examinations. Out of a class of 175 students, never less than 170 students have passed without the need for re-sitting examination.

Discussion

The basic principle of learning from practical exposure to the real world is not new. It is the basis of teaching by apprenticeship and it has been applied to the medical and other professions for many years. Originally, there was a flow of information and learning that went from seeing or assisting the professional persons in their work, to learning why the work was done in that manner, to finally learning the general principles involved. The learning process was slow but effective; the cry for relevance was not heard. There developed however, in the course of educational evolution a period of time when didactic learning was parrot-like, with drill sessions and the like. In response to this approach more effective ways of learning were designed and the students developed a better commitment to a life of continuing education.

The relevance of basic medical science to medical practice can be readily demonstrated (Tosteson, 1970) and one of the earliest course descriptions using patient histories was presented by Saffran (Saffran, 1971; Saffran and Franco-Saenz, 1975) in which the students gathered their own information at the bedside. For many programs this approach was impractical, since there was an increasing demand by the several different health professions for bedside teaching and the increasingly adverse reactions of patients to multiple examinations by students. Although it was perhaps a more theatrical approach than was necessary for the student to learn biochemistry, it was, nevertheless, a valuable demonstration of teaching biochemistry from case histories and it satisfied the continuing demand of medical students to see patients as early as possible in their medical education. Moving away from the bedside, but still presenting illustrative clinical material prior to the discussion of the component biochemical principles, was successful in stimulating student interest and it contributed to their understanding of the basic sciences (Rubinstein, 1972). The introduction of clinical examples into the lectures has, of course, been used most commonly to demonstrate the application of biochemical principles. In fact, several of these principles had derived from astute observations by clinicians who recognised that particular diseases might be the result of aberrant biochemical processes. Perhaps the most quoted examples are the metabolic pathways of glycogen metabolism and blood clotting with the cascade effects of their initiating stimuli.

The value of laboratory exercises is being increasingly questioned and debated (Spencer, 1979; Kogut, 1976) in the training of medical students. This is of particular concern with the development of associated medical science fields, such as medical technology, which remove the clinician further from the need to be skilled in basic science laboratory techniques. In many programs laboratory instruction has been replaced in the curriculum by problem-solving, case analysis, or examples of clinical simulation.

The use of clinical cases to teach biochemistry, sometimes combined with biochemical games (Carrington, 1978; Winkler, 1978) or clinical simulation (MacQuire and Solomon, 1971; Blanchaer, 1975), can take slightly different approaches to the learning experience. In some instances the information on the case is presented in parts, each part calling for a suitable response from the student before the next part is given (Smith and Jepson, 1972; Baggott and Trojak, 1978; Vella and Martin, 1975), and use has been made of audio-visual aids to do this (MacQueen et al., 1976). We chose to present all the facts together in the case, so that those of biochemical consequence could be grouped to formulate answers to questions that searched for the facts, the

principles and the associated problems of a biochemical nature
(Montgomery *et al.*, 1974). A similar approach has been taken by others
(Miller *et al.*, 1979). In all techniques the student learns to apply
biochemical knowledge to the analysis of appropriate medical
problems, the diagnosis being given in biochemical terms and the
treatment being seen to make biochemical sense. The student must
select from the case what data are needed and by a logical procedure of
scientific inquiry arrive at sound biochemical conclusions.

Implementation

The most frequent comment concerning the use of the case-oriented
approach to teach biochemistry is that most biochemistry teachers are
not clinicians. This implies that the approach involves more than
teaching biochemistry, which it does not, and that only clinicians can
discuss case material, which is clearly not true from our experience.
The concept at Iowa was introduced by the request to the department
that an experiment in teaching by the case-oriented approach be
conducted for four years by a group of four faculty members who were
interested in the idea and who were dissatisfied with earlier efforts in
teaching medical students. Three of the four were not clinicians.
Didactic lectures were given three times a week with a fourth clinical-
type lecture added frequently. A handbook of cases was prepared,
designed essentially as described above, with a brief 'language' section
in each of 13 areas of biochemistry. Each subject area had two cases
worked out as models and three or four cases for which model answers
were not developed. The class was divided into eight small groups, each
meeting with a faculty person for two hours each week. At these
sessions one of the model cases was reviewed by the faculty and a few
students then presented one of the cases that they had analysed. Each
student submitted a written analysis of an assigned case each week. A
short 15-minute examination was taken weekly covering the factual
information of the previous week's lectures and three major one-hour
examinations were given during the 15-week period of the course.
These examinations were cumulative and of a multiple-choice format,
using the style of questions that the student could expect to meet on
national or state board examinations.

The first year exposed a number of problems. In the small-group
discussion the students at first searched more for the physiologic or
pathologic aspects of the clinical material whereas the course was
intended to illustrate the application of biochemistry to health
problems. The students quickly assumed the role of doctor in writing

about the cases, some referring to the patient as 'my patient'. It was necessary to correct this attitude as soon as possible; nevertheless, the students were keen and they wanted to know all about the case. Not infrequently the faculty had to admit ignorance of these non-biochemical aspects, which would be covered later in the medical curriculum. The weekly written analyses of the students became increasingly long until a limit of space was prescribed for the answers. The last significant problem was the different manner whereby each faculty person dealt each week with the small-group discussions, a common core of material not having been arranged between them beforehand.

A significantly-revised handbook was written for the next class of medical students containing a new set of unworked cases. The written responses by the students to these case analyses were reduced in length. The discussion sessions were planned at a previous meeting of the faculty so that a common core of information to be covered was prepared each week by the faculty person lecturing on the didactic segment of the principles to be illustrated by the chosen cases. Minor revisions and more new cases were introduced into the third handbook which became the rough draft of the published text (Montgomery *et al.*, 1974).

Throughout these experimental years frequent and regular meetings were held with student representatives to discuss all aspects of the course. Their problems were resolved as quickly as possible and from the outset their attitude was most positive. After the fourth year of the course, by which time the more senior students were giving helpful feedback to the entering students, the faculty group was gradually replaced and at the time of this writing ten different faculty members have taught in the course, only two of whom were clinicians. Others have agreed to teach in the future and those who prefer not to engage in teaching the course either do not teach any health professional courses or prefer not to teach in courses that have four or five faculty involved. The comment that most biochemistry teachers are not clinicians is not particularly relevant to case-oriented courses. Obviously additional preparation is required for those faculty involved for the first time, as is true of any new teaching experiences. Thus the case-oriented approach primarily calls for a commitment to teach students how to *use* biochemistry in their profession.

Results

The student response to teaching from case material has been increasingly positive. No less biochemistry is taught than was presented in earlier courses; in fact the course objectives for medical biochemistry have been expanded to include that material required to analyse the clinical material. In particular, this additional material includes nutrition, fluid and electrolyte regulation and molecular endocrinology. After completing the course students have been better prepared to study the later courses in the basic sciences, such as physiology and pharmacology, and comments from our clinical faculty, who have been very supportive of our efforts, indicate an overall improvement in the appreciation and understanding of the basic mechanisms of disease. Student evaluations of the courses were conducted each year, but except for subjective expressions of support for the approach, no quantitative analysis of comparison with earlier teaching methods was made.

Other courses in biochemistry for the health professions have selected appropriate material from our text on the case-oriented approach, physicians' assistants having been taught at Iowa in this way since 1973 and occasionally dental students. The results with these students again are claimed to be successful in stimulating them to use the biochemical information for analysis of appropriate diseases or clinical situations.

The case-oriented approach has been extended to other subjects (Barnett, 1976), and it is perhaps coming full circle in its application to the teaching of the fundamentals of all subjects. Two times two equals four is academic unless its application is understood.

Acknowledgements

I am deeply indebted to my colleagues, R. L. Dryer, T. C. Conway and A. A. Spector, who joined me in the early development of the case-oriented approach to teaching biochemistry and shared the trials and the satisfaction of publishing our results.

4 INNOVATIONS IN THE TEACHING OF MEDICAL BIOCHEMISTRY

F. Vella

> The essence of knowledge is, having it, to apply it.
> Confucius (551–479 BC)

In reviewing innovations that have occurred in the teaching of medical biochemistry in the past quarter century, I am struck by the multitude of factors that have produced, or that motivated, such innovations. Among these factors could be listed:

(1) the application of the findings of educational psychologists;

(2) the development of new philosophies and strategies to encourage learning by students;

(3) the demands of students brought up largely in educational institutions based on the previous two factors, and who seek interest, relevance and individual consideration;

(4) the development of educational technology (such as computers, individual learning through tapes, reading assignments, films, slides, student-conducted experiments guided by tapes and/or printed instruction sheets);

(5) the vast expansion in the knowledge and understanding of the molecular basis of living systems and especially of the human organism and of disease; and

(6) the many cultural and economic changes that have occurred during the period.

It is my belief that similar factors will continue to play a significant part in the evolution of teaching–learning philosophies and practices of medical biochemistry.

Innovations are really non-conventional ways of doing things. The various innovations that are the subject of this chapter are non-conventional in that they are squarely based on the principles that learning requires the active participation and involvement of the learner, and that the learner is eager, willing and capable of learning for himself (or herself) if informed of what it is he is required to learn. They are therefore more concerned with learning processes than they

81

are with the details of the subject matter that is to be learned. They were initiated by teachers who have been able to redefine their role from that of 'transmitters of specialised information' to that of guides and counsellors of those who want to learn biochemistry as a basis for, and as related to, the study of medicine. This contrasts with the conventional method of teaching biochemistry through formal lectures and practical laboratory exercises only.

It has been said many times that biochemistry is a difficult subject not only for the learner to learn but also for the teacher to teach. This opinion seems to know no geographical boundaries (Campbell, 1975a; Gray, 1975; Hall, 1973; Rubinstein, 1972; Sable, 1975) and is by no means new (Acheson, 1954). Mehler (1973) has expressed the 'conviction that in general the teaching of biochemistry in professional schools has been a total failure' and similar views have been echoed by others (Alexander, 1973; Campbell, 1974; Evered, 1975; Kogut, 1974; McIntyre, 1975). Chapman (1979) has taken the extreme view that 'education for medicine in its pre-medical and pre-clinical phases is intellectually deficient, horrendously wasteful in money and in time and in urgent need of overhaul'. This pessimistic picture is to be balanced, however, by the many attempts that have been made (with a good measure of success, let it be said) to change the emphasis on formal conventional teaching with its production-line method into one of learning by students who differ in styles, backgrounds and interests. These attempts have been characterised by the variety of approaches that have been used to encourage student interest and involvement in a 'living chemistry', i.e. biochemistry, as it relates to the human organism in health and in disease.

Some of the innovations I am concerned with here, have already been reported on extensively and will be presented here briefly. Others merit slightly longer consideration since they may not be widely known. A description of the present status of practical laboratory courses, leads to a review of methods designed mainly to enhance problem-solving skills (case studies, structured learning experiences, Jepsons, card games) and of methods of integrated teaching and independent study (personalised system of instruction, audiotutorial exercises and computer-assisted instruction).

Practical Laboratory Courses

Dissatisfaction with laboratory courses for medical students is not a recent development (Acheson, 1954). Though a significant number of departments have discarded laboratory exercises completely, many have been reluctant to do so because of the significant role they are supposed to play in medical education. A survey of 64 departments in North America (Guyer *et al.*, 1975) found that laboratory teaching had been halved during the previous four years (from an average of 94 hours to one of 47 hours with a range of 0–120 hours) and that two-thirds of these departments used traditional laboratory assignments for at least half of their laboratory time. The two objectives for laboratory assignments which were most highly rated in this survey were 'supplementation and reinforcement of didactic material' and 'appreciation of experimental development and methodology'. A more recent survey (Devlin, 1979) has shown a decline (to 45 per cent) in the number of departments which present a laboratory course and in the average time allocation to such assignments (to an average of 30 hours, range 2–92 hours). This marked trend to reduce laboratory time and to replace traditional exercises by problem–analysis, independent study, clinical-biochemical conferences, demonstrations, student projects etc. is also true outside North America (Campbell, 1975b; Laburn *et al.*, 1978).

The virtues of practical courses in biochemistry have been defended by several authors (Dixon, 1978; Karlson, 1977; Spencer, 1979). However, the need for alternatives in which the use of complex apparatus or laboratory procedures are demonstrated and emphasis is placed on such skills as the acquisition of information from various sources (e.g. tables, figures, diagrams), the interpretation of results of experiments and the design of experiments to investigate specific questions has been well recognised (Campbell, 1975b; Kogut, 1976; Landon and Mayer, 1976; Mehler, 1973; Vella and Martin, 1976).

Problem-solving Skills

There has been a great amount of interest in the development of the problem-solving abilities of medical students. In biochemistry this has led to the use of problem-sessions often as alternatives to laboratory exercises. Many of these sessions appear to be based on theoretical problems and often on the acquisition of mathematical or laboratory skills as exemplified by the problems presented at the end of each chapter in Lehninger (1975) or Bohinski (1979) rather than on the three activities of observation, analysis and deduction which are basic to problem-solving (Hyzer, 1977).

The general ability of problem-solving seems to be innate in some individuals. Most of us, however, have to develop and integrate these separate abilities through experience or training. It has been categorically stated that 'many students enter University unable to reason – to think logically and independently, to observe and infer, to analyse, criticise and explain – and some of them depart with this disability intact' (Hartman, 1978). This sweeping generalisation ignores the reasoning ability which is necessary for everyday life which all intelligent persons have to a significant degree and presumably refers to the reasoning involved in academic pursuits. This is not difficult to understand. Problem-solving entails the hardest kind of mental effort as problems perplex, challenge and demand confidence in our ability to solve them. De Bono (1970) has defined a problem as 'simply the difference between what one has and what one wants' and problem-solving as 'the process which is used to bring about a change in this state of affairs'. Training in medical biochemistry should therefore be concerned with directing the general problem-solving abilities of students toward clinical-biochemical problems.

Didactic experiences intended to enhance these abilities include presentation of patients, of clinical histories and of experimental studies which relate to disease processes, clinical simulation ('Jepsons') and association games (card games) and written assignments based on research publications (structured learning experiences). Blanchaer (1975) has described the use of written simulated clinical problems which involve the making, at various stages, of diagnostic and patient management decisions.

Case Studies

Many students find biochemistry difficult to learn and appreciate the subject much more when care is taken to relate it to clinical problems. The presentation of clinical problems motivates many to learn the basic biochemistry that is needed to understand, solve or manage such problems, and reinforces that knowledge by extending it into the realm of practical applications. As a result, a strong tendency has developed to orient biochemistry courses towards clinical problems. One textbook has adopted the 'case-oriented approach' (Montgomery *et al.*, 1977), while another is based on the premise that 'human biochemistry is studied in order to relate it to human disease' (Pasternak, 1979).

Attempts have been made to assess the biochemical content of clinical practice in an effort to offer guidance as to which areas of biochemistry are most often utilised clinically. Wills (1974) found that

38 per cent of the articles published in a four-year period in the journal *The Lancet* included 'an important biochemical topic'. Hormones, the immune system, electrolytes and trace metals, and nutrition were cited most frequently (58 per cent of the total citations) in the 'biochemical' articles, while metabolism accounted for only 3.4 per cent. These broad findings were largely upheld by a study of the activities of a group of clinicians and of the relation of their activities to biochemistry (Kogut and Cramp, 1975). However, the debate as to what and how much (if any) biochemistry a doctor needs to know continues unabated (Candlish, 1974; Kornberg, 1978; Newsholme, 1979; Schwartz, 1979).

Case studies have been used extensively in basic biochemistry courses in several medical schools (Montgomery *et al.*, 1977; Rubinstein, 1972; Saffran, 1971 and 1973; Vella and Martin, 1975 and 1976), while 'clinical topics' are discussed extensively in others (Dakshinamurti, 1979). They offer many positive features:

(1) they present real problems whose significance in terms of human experience is easily recognised and appreciated by students;

(2) they demonstrate how clinical problems have been defined, analysed and managed, and the outcomes of such definitions, analyses and managements;

(3) they utilise and extend the student's stock of knowledge;

(4) they can be varied in scope, complexity and frequency of occurrence of the problem and can be enriched with reviews of experimental research on the topic;

(5) they demonstrate the unity of knowledge by integrating biochemistry with the many other 'disciplines' which relate to the human organism and its activities;

(6) they are good entry-points for many areas of biochemical knowledge;

(7) they constitute problems about which students may never have heard;

(8) they are enjoyed by students;

(9) they can be used in a variety of ways, e.g. to introduce one or more lectures on a biochemical topic, to demonstrate how biochemical knowledge is applied, as problems for assignments, tutorials, self-instruction etc.;

(10) their use recognises that the major objective of medical students is the acquisition of the skills needed for medical

practice rather than of biochemistry as a discipline *per se.*

Significant negative aspects are that first they require that the instructor of biochemistry have some understanding of, and familiarity with, human pathophysiology and that he be prepared to discuss clinical problems and to read clinical as well as biochemical literature, and secondly they require that biochemical knowledge be seen as a means to an end in the study of medicine rather than as a subject which is very satisfying in itself.

Structured Learning Experiences (SLEs)

The teaching of biochemistry should provide not only the theoretical information which is needed for the definition and possible solution of clinical problems but also the ability to appreciate and to understand new developments as they are reported in the scientific literature. Since 20–40 per cent of all articles published in journals such as *The Lancet* and *The New England Journal of Medicine* deal with problems having a chemical or biochemical background (Astrup, 1975; Wills, 1974) and there is no sign that these figures are decreasing, students should be trained in critical reading of the literature.

Essay writing is the traditional way of introducing students to the scientific literature. This has often resulted in verbatim quotations from abstracts and discussions of papers, rather than in analysis of the design of the experiments, the results obtained and their significance. Detailed study of one published paper with emphasis on knowledge of terminology and the theoretical background, the reasons for and design of the study or experiment, the techniques used and the results obtained was felt to be a more direct way of guiding the student through the literature and is the principle which underlies SLEs (Vella and Martin, 1976). Similar approaches have been reported (Baggott and Trojak, 1978; Bender, 1977; Frunder, 1978; Spencer, 1979).

The subject-matter of an SLE is a published paper selected because of its human biochemistry interest. It is rewritten by the instructor so as to describe the problem defined by the author (or authors), the objective of the study, the design of the experiments which were undertaken and the results obtained. The material is presented in a logical sequence under the following headings: Principle (this briefly describes the state of knowledge at the time of the study), Results of Experiments, Summary and Conclusions. The results are presented in table form and as reproductions of illustrations in the original paper. Each section is followed by a series of questions which relate to theoretical aspects, understanding of the experiment, analysis of the results obtained, consistency between results of different experiments in the study, interpretation of results and development of hypotheses to explain the results. The questions are the core of the SLE and may be of various categories (descriptive, interpretive, integrative, summative etc.) and levels of complexity. Written answers to these questions are handed in for grading and comments by the instructor. Students can work through an SLE as individuals or in small groups during the time scheduled for laboratory work and can consult an instructor as they see fit. SLEs emphasise concentration on the meaning of written material, accuracy of thought, problem-solving and the need for perseverance.

SLEs have many positive features:

(1) they are concerned with procedures used in the acquisition of new biochemical knowledge as well as with an area of biochemical knowledge;

(2) they involve students through the use of higher cognitive processes concerned in problem-solving;

(3) they emphasise guided self-instruction over passive learning from an instructor;

(4) they permit synthesis of new, or of previously-learned, information, concepts and principles from a variety of disciplines;

(5) they encourage students to work on a problem rather than to turn in an answer;

(6) they present experimental procedures and techniques not normally available in undergraduate laboratories because of expense, or lack of specialised equipment or of expertise, and simulate current laboratory practices;

(7) they are recognised by students as a worthwhile intellectual challenge which is stimulating and often enjoyable;

(8) they encourage consultation between students and instructor;

(9) they may be used in different ways (for example, as classroom or as home assignments, as individual or as group efforts) and with different objectives (for example, with emphasis on laboratory techniques, or design of experiments);

(10) the material covered seems to be better remembered and more easily applied to new problems since the learner has actively to manipulate or restructure his knowledge.

The negative aspects of SLEs include:

(1) students who have been involved mostly with traditional forms of education, find this type of problem-based learning difficult at first, though they soon adapt to it;

(2) students may have to be taught intellectual procedures that may not be generally considered part of the repertoire of instruction in biochemistry (for example, problem formulation, problem summarisation, interpretation of illustrations etc.);

(3) heavy demands are placed on the instructor especially in the preparation of SLEs.

'Jepsons' (Information Games)

'Jepson' is the name used by Vella and Martin (1976) for a paper exercise in which groups of up to five students strive to determine the molecular basis of a patient's clinical problem from written information supplied to them. These games differ only in detail from those described by Smith and Jepson (1972).

Information that is presented comprises a brief case history, results of routine laboratory investigations (with normal values for comparison), a family history if appropriate, and results of special investigations usually of a research nature, presented on separate sheets of paper. Each game consists of five to nine sheets. A logical sequence is used in the presentation of each sheet to each group, and the last sheet usually contains the results which identify the molecular lesion.

These games are best undertaken over two-hour periods, of which the first 50 minutes or so are taken up by the game proper and are followed, after a pause for rest, by a critical discussion by the instructor of the information presented and of the significance of the molecular abnormality involved. Use of these games has been reported not only with classes of medical students (MacQueen *et al.*, 1976; Miller *et al.*, 1979; Randle, 1975; Winkler, 1978) but also in science courses in biochemistry (Peacock and Tribe, 1979; Suckling *et al.*, 1979).

The advantages that have been listed for case studies apply equally to these games. Besides, Jepsons are easy to produce, involve students actively in problem-solving and the manipulation and restructuring of knowledge, and enhance the skill of 'problem distillation' whereby an accurate, ongoing, mental summary of a problem is made as it unfolds for the participants. They supply an inner reward for the accuracy of reading and thinking and the pooling of resources that are involved, through the satisfaction that inevitably arises from correct identification of the molecular lesion.

Card Games

Carrington (1978) described two card games ('Pairs' and 'Bingo') the objective of which was to increase the exposure of students to the names and molecular structures of biomolecules. Card games can also be used to encourage, and to develop, the making of associations between items of information written on card faces. One such game ('CVS Rummy') has been devised for use by medical students as a physiology laboratory exercise (France, 1979). I have applied this idea to 'Biochemistry and Disease' and 'Biochemistry of Vitamins' and have used these games as laboratory exercises in medical biochemistry.

Each card game consists of a pack of 52 plain index cards on one face of which is written a word (or phrase) or a molecular structure (with or without its trivial or chemical name). The information on the cards is such that they can be organised into 13 sets (each of four cards) and each set relates to a particular disease (or vitamin) and appropriate aspects of its biochemistry. Other combinations are also possible. Coverage of subject matter can be extended to intermediates in metabolic pathways, structure and function relationships, drugs and antimetabolites, constituents of enzyme reactions, inborn errors, nutritional deficiencies, hormonal disorders etc.

Such games have been played in small-group or individual modes. The only rule is that each set must be justifiable to an instructor. Very useful for a starter, is the 'solitaire' format in which cards are exposed one at a time with grouping and regrouping until the maximum number of sets has been accumulated. If desired, a competitive element can be introduced by having groups work against the clock or against each other. Repetition of games consolidates knowledge and associations, and can utilise a 'Rummy' or other, format. A game session starts with a very brief description of the objective of the game and ends with distribution of copies of the information grouped into sets, and follow up by the instructor in which questions are answered and the rationale for each grouping outlined.

Students have received these games enthusiastically. Their content was perceived as relevant. The exercises were acknowledged as good practice in the use of biochemical information and for establishing relationships between items presented in a disconnected fashion.

A good card game engages the sustained attention of learners, reinforces learning in each game cycle, entails learner initiative, requires appropriate use of information besides mere recall, and induces satisfaction rather than boredom or fatigue (Short, 1979). These games seem to satisfy these characteristics. Besides, they are not difficult to devise and can be used in a variety of ways (for example, in tutorial sessions, for evaluation or review etc.). They are also inexpensive and require no special facilities or expertise on the part of instructors.

Integrated Teaching and Independent Study

The breakdown of departmental boundaries in teaching at medical schools started with the birth of the 'integrated curriculum' at the medical school of Case Western Reserve University at Cleveland, Ohio almost 30 years ago. This approach has been adopted by over one-quarter of the medical schools of North America (Thompson Bowles, 1974). Integration of biochemistry 'horizontally' with cell and molecular biology (Schepartz, 1974; Campbell, 1975a) and 'vertically' with medicine and paediatrics (Miller *et al.*, 1972) have been described. The latter authors were 'surprised by students' ability to grasp, intuitively and immediately the ultimate vitally inter-dependent relationship between basic science and clinical medicine'.

There is little doubt that learning of biochemistry can be approached from a large number of entry points and can be adapted to the learning styles of individual students so as to take advantage of differences in their motivation and backgrounds. These ideas have been incorporated to the greatest extent in the McMaster curriculum (at Hamilton, Canada) which is based completely on self-directed, problem-based, small-group learning (Neufeld and Barrows, 1974; Neufeld, 1974). Several new medical schools in the USA. (Bloomfield *et al.*, 1973), Britain (Editorial, 1970), the Netherlands (Cuypers, 1978) and Australia (Kellerman, 1978; Clarke, 1978) have utilised similar principles. In these curricula, learning objectives and the level of mastery are largely prescribed by faculty and the materials used include computer-aided instruction, programmed texts, slide/tape material, printed texts, video cassettes and laboratory exercises, but formal lecturing plays a minor, if any, role.

The principle of independent study has been adopted, to varying extents, also by older medical schools. Of 90 medical schools surveyed in the USA, 34 per cent had independent-study programs and most of them were in basic sciences (Trzebiatowski, 1976). The same survey revealed that 45 per cent of schools that did not have them planned to develop such programs in basic science subjects within two years. The movement towards these approaches is, therefore, strong.

These innovations have been accompanied by sometimes heated discussion as to who is best qualified to teach biochemistry. In Britain, where biochemistry has been described as the 'neoanatomy' of the medical curriculum, the most vociferous have claimed that the subject should be taught by the clinical biochemist (or chemical pathologist) (Goldberg *et al.*, 1970; Whitby, 1974; Gray, 1975) especially since 'very few biochemistry teachers have much knowledge of human as opposed to animal biology and few still are medically qualified' (Sinclair, 1975). This view has not been widely accepted. Already, in many schools, clinical biochemists assume responsibility for case presentations and clinicopathological conferences that show how biochemistry is used in medicine (Goldberg *et al.*, 1970; Schwartz, 1975) while, occasionally, biochemists have found a role in teaching 'bedside biochemistry' (Saffran and Franco-Saenz, 1975).

Newsholme (1979) has singled out what is probably the most important of the difficulties that underlie the adoption of these, and other non-conventional methods of teaching, when he wrote:

> What is needed is not necessarily a medically qualified person with an interest in biochemistry, but a biochemist, medically qualified or not, interested and prepared to work sufficiently hard to get to know his subject and the physiological and clinical applications.

Personalised System of Instruction (PSI, Keller Plan)

In a course organised on the Keller plan, students work on their own, at their own pace within a time period, in a mastery-oriented program in which the course material has been divided into units of convenient size, each delineated by carefully-written objectives, with specification or provision of texts which contain the material required to be learned. Mastery is assessed on completion of each unit and is required before a student can proceed to the next. Few, if any, lectures are given and the role of lecturer is changed dramatically to that of program-developer, tutor and counsellor. The method was developed by Fred Keller (1968) and is suitable for classes with a maximum of about 120 students per instructor.

Practical advice on the setting up of Keller-type courses, their advantages and disadvantages have been well-described (Elton *et al.*, 1973; Kulik *et al.*, 1974; Stoward, 1976). A substantial amount of experience with such courses in biochemistry for medical and for science students has been accumulated in the USA (Weisman and Shapiro, 1973; Cohen *et al.*, 1973; Pearlmutter and Pearlmutter, 1977; Calvo, 1978), Argentina (Batlle, 1975) and Spain (Galindo *et al.*, 1976).

Audio-tutorial Exercises

Experiences with the use of tape-recorded commentaries to guide the student through a sequence of learning exercises which may include slides, displays, printed texts, experiments and problems were described over a decade ago in the teaching of electrocardiology (Owen *et al.*, 1965), biochemistry (Garcia *et al.*, 1968), endocrinology (Harden *et al.*, 1969), obstetrics and gynaecology (Chez and O'Gorman, 1970) and in botany and zoology (Postlethwait *et al.*, 1971). The method has been extensively developed in the Department of Biochemistry at the University of Dundee where it includes simple experiments and tests of the type a clinician might have to perform for himself or which illustrate a general principle such as enzyme assay or spectrophotometry (MacQueen *et al.*, 1976). It has also been employed in biochemistry courses for science students (MacQueen *et al.*, 1976) and at the postgraduate and postdoctoral levels (Garland, 1975).

The effectiveness of slide–tape or slide–print instructional materials as compared to lectures and practical laboratory exercises has been amply recorded (see Halcomb and Garner, 1973 for references). Such materials have been widely used and accepted as an adjunct to, or a replacement for, the conventional methods. A well-developed lecture script can quickly and easily be converted into a slide/tape unit of instruction which is educationally effective and is well-received by students (Baggott *et al.*, 1977; Tomlinson, 1979). The advantages and disadvantages of this method have been described (Halcomb and Garner, 1973; MacQueen *et al.*, 1976).

Computer-aided Instruction (CAI)

The use of computers for dissemination of information in various courses in the formal education of medical students is well over a decade old. It was largely pioneered in the Independent Study Program of Ohio State University College of Medicine (Prior *et al.*, 1970; Folk *et al.*, 1976). This program has been successfully adapted by two other medical schools in the USA (Trzebiatowski, 1976).

CAI is in use in teaching medical biochemistry, to varying extents, in several schools in North America (AAMC, 1975–6). Its use should increase rapidly, especially where utilisation of available resources and facilities may only require the installation of terminals and of telephone communication lines. The use of computers for education at all levels promises exciting vistas for the not-too-distant future as computer technology becomes more and more adapted to the service of the teaching–learning process and attains the capability of providing computers that serve as personal assistants for learning (Goldstein and Brown, 1979).

There are at least three uses for computers in education: (1) to present instruction directly to the student and to assist, or substitute for, the instructor, (2) for simulation, demonstrations and gaming, and (3) for evaluation of student mastery of knowledge and or problem-solving skills. Removal of the limitations of frame-based CAI, use of techniques of information-processing psychology to build models of learner skills and replacement of computer jargon by natural-language capabilities will enhance these roles remarkably. Much excitement in the role of computers in learning arises from the fact that they can serve in ways that books or television cannot, since learner and computer can interact in many ways which make the computer more captivating, more challenging and more involving than many other methods (Goldstein and Brown, 1979).

The basic philosophy and methods of CAI have been well-described and reviewed (Stolurow *et al.*, 1970; Halcomb and Garner, 1973; Deland, 1978).

Concluding Remarks

This chapter has described some of the innovative changes that have occurred in the teaching of medical biochemistry in the past quarter century. Though pessimism has been expressed about the prospects for change in teaching of basic sciences in medical schools (Jacobson, 1971), and a return to rather traditional medical education models has been detected (Fogel, 1976), it is safe to say that many efforts are still in progress in the attempt to decrease, and if possible eliminate, the sharp division that still exists between basic sciences and clinical sciences.

Where learning is still equated with the number of hours of student exposure to faculty (Thompson Bowles, 1974) and where it is more convenient to teach biochemistry (or any other subject) in the passive-assimilation-of-knowledge mode, then change is threatening and unwelcome no matter how reasoned the arguments for it may be. Where it is believed that the practice of medicine is a kind of problem-solving and that problem-solving is another name for learning (Tosteson, 1979), then innovative approaches and self-instruction modes thrive.

Apathy, and in many instances resistance, is the response of faculty members towards the adoption and use of new educational approaches and strategies in medical biochemistry as in other subjects (Ashby and Besse, 1972; Vella, 1977). Teaching is a very personal matter and a teacher's approach is determined by his or her understanding and appreciation of its objectives. If knowledge is seen as the acquisition of information, the teacher will see his role mainly in information transfer. If the utilisation of information is seen as a more important aspect of knowledge, the teacher will see his role as that of developer of the student's innate logical and reasoning skills. My own preference is for the humanistic view that has been expressed as follows:

We must acknowledge again that the most important, indeed the only thing we have to offer our students is ourselves. Everything else they can read in a book or discover independently, usually with a better understanding than our efforts can convey. (Tosteson, 1979.)

5 THE USES AND ABUSES OF ASSESSMENT IN BIOCHEMISTRY EDUCATION

D. Rex Billington

Introduction

There has always been assessment in biochemistry education, but different words have been in vogue at different times to describe this activity — examinations, quizzes, tests, measurements and evaluation. Today the 'in-word' is assessment. Because it is impractical in one chapter to discuss all aspects of assessment in biochemistry education, this discourse will be selective and limited to an overview of the uses and abuses of assessment. Particular attention will be paid to the purposes of assessment and some of the newer techniques which have appeared on the biochemistry education scene.

The fundamental purpose of education is to affect changes in the way an individual thinks, feels and acts. The process of educational assessment is one of gathering and fashioning educational data into an interpretable form so that decisions may be made about the effectiveness of an educational process in bringing about these desired changes. This applies in biochemistry education as in any other part of education. Because education is about thinking, feeling and acting, education tests are classified according to these three dimensions. Thinking tests are usually called cognitive, and sometimes inappropriately called knowledge tests. Cognitive tests vary in degree from very simple tests of knowledge recall to complex tests of original and creative thought. Multiple choice, matching, fill-in-the-blank, true/false and essays are examples of traditional formats used to assess thinking and academic achievement in the biochemistry classroom. Pencil and paper problem-solving tests are now appearing more frequently. Feeling tests are also called affective, non-cognitive, or attitude tests. They are used to assess interest, attitudes, appreciations, values and emotional sets. They vary from very simple yes/no inventories such as might be used to find out whether students like a particular film or not, to complicated personality tests of character and conscience. Feeling tests should not be used to award student grades. Cognitive questions have the same correct answers for all respondents and therefore are gradable, whereas the correctness of an affective question depends purely upon the person queried, there being no absolute or consensual right or wrong. Affective tests are useful in the assessment of innovation and for guidance purposes. Acting tests are usually called psychomotor or skill tests. They require the manipulation of materials and objects using our sensory and neuromuscular systems. Some aspects of biochemistry laboratory and clinical assessment fall into this category. Most biochemistry laboratory and clinical tests, while usually intended for measuring either cognitive or psychomotor skills, do reflect some attitudinal or affective elements in them.

It is not difficult to justify discussion of cognitive and skill assessments in a review of this sort. However, there may be some hesitancy about including non-cognitive assessment in such deliberations because it is often regarded as less objective, less empirical, and perhaps more peripheral to academic activity. But it is unlikely that one does much thinking or acting without feeling; a person usually responds as a total organism. It is sometimes said that students perform best in subjects they like and that students who perform best learn to like the subject as a consequence of the success it brings them. Both points of view are probably true and not in conflict. Because we cannot divorce feeling and learning we must consider non-cognitive assessment in any educational review of biochemistry, although we must also pay considerable attention to the quality and limitations of this measurement.

In this chapter there will be no discussion pertaining to assessment in the scientific field of biochemistry. Whenever the terms assessment or biochemistry assessment are used, they refer to biochemistry education assessment. This does not deny that many of the principles which are fundamental to educational evaluation have their counterparts in biochemical assays and research. It should also be noted that biochemistry education as referred to in this chapter relates to higher education as separate from primary or secondary education, though it is acknowledged that biochemistry education goes on in secondary schools.

Purposes of Assessment

Tyler and White (1979) describe a wide range of purposes for assessment in education:

> In the classroom, some teachers are using tests to make assessments in order to individualise instruction. Some are using tests as means for placing students in a sequence of an educational programme.

Some monitor the progress of individual students by periodic testing. Individual students are taking mastery tests to demonstrate their successful completion of a unit of instruction. Tests are being used by many teachers to diagnose the learning problems of individual students who are having difficulty.

Curriculum builders are using tests for the assessment of group needs. Administrators are using tests to monitor the progress of classes in schools. Administrators are using tests to assess student achievement at certain age or grade levels. Tests are being used to evaluate educational programmes. Tests are also being used by various persons to appraise the educational effectiveness of instructional equipment and material. Tests are being used by guidance personnel to furnish information to the student that is helpful in his planning for his future. Tests are also being used by legislators and other officials at various levels of government to obtain information about the progress and problems of education in their jurisdiction. Tests are also furnishing information to the general public about the progress of education.

Biochemistry teachers are using assessment for some of the purposes described above, though Tyler and White were referring to the use of tests across the total education scene, with particular reference to the USA. One may surmise that there are many purposes for assessment above and beyond giving students marks for work done.

Influences on Assessment Procedures

What are the influences on the selection of assessment procedures in biochemistry education? Two major influences are the purposes of education and teaching methodologies. Until relatively recently a dominant use of primary and secondary school assessment was to grade, classify and sort people for the job market and for the limited number of places in higher education. Although a little less clearly, assessment in higher education probably served largely to legitimise the awarding of degrees and diplomas and help select students for employment or still more higher education. In all sectors of education, assessment was essentially a discriminating and competitive activity sometimes culminating in the posting of ranks in class. But there is now a gradual change in the purposes of assessment because there is a gradual change in the purposes of education. One of the forces of change relates to employment. With the decline in the need for unskilled labour due to the impact of technology, people without a good education are having particular difficulty in finding work. Unemployment places a considerable burden on the state especially in modern societies such as Britain. The social demand for education is still strong despite the financial difficulties facing state-financed systems. It is generally agreed that young people need to be prepared, through education, to meet the demands of modern society whilst older people need to be encouraged to adapt. The critical task of schools and education systems now is no longer one of sorting out for the employment market, it is as much, perhaps even more, one of sorting in. More schools, colleges, polytechnics and universities now encourage people to stay in education longer. Assessment based on inter-student competition, which is oriented to ranking and intended to discriminate among students, is being replaced by assessment which provides private guidance to students on their personal progress and mastery of the objectives set for them. This latter form of assessment is more constructive for the purposes of keeping students in school. There are other sound pedagogical reasons for this approach to assessment also, but these will not be immediately pursued.

This mastery testing, as it is sometimes called, is becoming a distinct feature of progressive medical and dental professional education programs. The reason for mastery testing here is perhaps not so much one of wanting to keep students in school longer, but more of wanting to minimise student attrition from expensive educational exercises without having to compromise professional standards. This again is another familiar example to biochemistry teachers of educational purposes and pressures influencing assessment procedures. Such influence is a normal state of affairs and should not be regarded as interference but as being part of the trend towards accountability in higher education.

Not only do the purposes but also the methods of education shape assessment. The effects on biochemistry education of the now not-so-new technologies of video-tapes, audio-tapes, and computers, are documented elsewhere in this book. These technologies have stimulated more varied, and have resulted in more valid, methods of assessment. Video brings to biochemistry assessment a range of aural and moving, coloured, visual stimuli, which may include detailed electron microscopic pictures of DNA, to panoramic views of civilisations decimated by radioactive fallout. The same audio-visual stimuli and test question may be presented by video to different students in different locations at different times. Computers provide the capability of storing or banking test questions (Morgan, 1979a; Buckley-Sharp and

Harris, 1970; Bryce, 1979a) of assembling individually-tailored tests for students, and for providing immediate and full feedback of test results (Haywood and Wood, 1977). This feedback may include detailed information not only to the student but also to the teacher about the strengths and weaknesses of his test using item-analysis techniques. When computer and video are linked, assessment can be very creative. Technology is helping to make assessment a more sensible, natural and positive part of the learning process and above all is helping to increase the validity of assessment. The pervasiveness of technology in educational assessment in biochemistry, however, is a function of educational affluence, for equipment and programming (software) is expensive. One may speculate on the potentially exciting prospects of microelectronics on assessment techniques in biochemistry education. But it does seem fair to say that the realisation of the prospects of this technology will be heavily dependent upon fiscal resources.

Uses and Abuses of Assessment

Preliminaries

The purposes of assessment in biochemistry education may be divided into four categories:

(1) to help in academic and career guidance and counselling for students;
(2) to help provide guidance to the teacher about his teaching;
(3) to help evaluate educational innovation and experimentation;
(4) to help hold teachers, educational institutions and educational systems accountable.

Before discussing the uses and abuses of assessment as they pertain to each of these four sets of purposes, the following three observations are presented to provide perspective. Some of the discussion is somewhat abstract and perhaps new to many biochemistry teachers. But the discussion is none the less important, because the consequences of assessment are important. One must know clearly why one assesses and have considered the potential impact of assessment in human terms. It is intended that the following discussions should stimulate constructive examination of current assessment practices. The second observation is to comment briefly on the concept of validity which is of fundamental importance to all measurement, but of special importance to mental and education measurement. A test is valid when it assesses what is intended to be measured. There is no such thing as a valid test *per se*, for a test is only valid in reference to a particular use. There are different ways of establishing and finding out whether our biochemistry education tests are valid or not. Without going into epistemological discourse which underpins educational assessment and arguments of validity, one may state with little fear of contradiction that there are two main types of evidence which help establish the validity of an assessment. They are, rational evidence and empirical evidence. Most validation exercises in education assessment use both types of evidence, but because of convenience greater reliance is usually put on rational than empirical. It is probably true to say, however, that empirical evidence impresses and would be more acceptable to biochemists and physical scientists than rational evidence. Much of the discussion in the following four sub-sections of the purposes of assessment pertains to issues of validity. The third observation to note, when reading the rest of this chapter, is that discussion of a particular assessment issue may apply to more than one of the four sub-divisions of purposes upon which the rest of the chapter is organised. For example, an issue such as pre- and post-testing which is discussed in relation to guiding students may also apply to assessment which is part of program evaluation.

Concerning Assessment in Student Guidance

Tests are used in the selection of students and selection by students of academic and vocational tracks. Assessment thus affects the career and progress of biochemistry students in a number of ways and at different times: results of secondary-school examinations help decide eligibility for higher education, eligibility for subjects in the first year and eligibility for certain professional vocational courses. First-level or first-year course performance helps guide decisions about suitability and eligibility of a candidate for second- and upper-level biochemistry courses. And subjects studied and successes gained in biochemistry examinations affect job opportunities and employment decisions. Guidance of students is partly based on academic eligibility rules, but there is a factor of choice for those students who have satisfied these requirements to decide whether they wish to pursue a particular vocation or biochemistry course or not. Examination results influence these choices.

Eligibility for Biochemistry Courses. How appropriate and valid are the criteria which decide eligibility for biochemistry classes? Chemistry, physics, mathematics and biology are the usual secondary-school science prerequisites for biochemistry courses in Britain. Some professions such as medicine and dentistry require very high passes in these science courses for entry into their degree programs. There is more flexibility in entrance requirements for some courses in biochemistry where students are preparing for jobs and professions in which biochemistry is less central and important. Biochemistry academics in Britain influence the objectives of secondary education and its assessment in a direct manner when they serve on national secondary-education curriculum examination committees. These academics also affect secondary education and its assessment indirectly when higher-education curricula influence what some secondary-school teachers teach and hence test their better senior-science students. But what of the relationship between high-school success and in particular first-year biochemistry success? Unfortunately the literature is replete in reporting good empirical studies which have examined the relationship between secondary-school academic performance in various subjects, science in particular, and higher-education biochemistry course success. Deciding school prerequisites for biochemistry in higher education is largely based on rational evidence. Those who are inclined towards and achieve well in science at secondary school would logically do best in higher-education biochemistry it is argued. Perhaps the conventional wisdom should be augmented by some empirical predictive research. The unavailability of good empirical predictive validation studies however, is some testimony to the difficulty of carrying out such investigations. One particular difficulty in studying this problem is that many students who do not achieve well in high-school science do not take biochemistry in higher education, so consequently do not figure in the statistics. Correlations between the two assessments therefore, would be based on a relatively homogeneous sample, hence a spuriously low correlation coefficient would be likely. Such low correlations may mislead some people into concluding that a poor relationship exists when that may not be the case.

There might be some value in looking into other predictors of success in biochemistry other than the secondary-school science subjects. It is not unusual to find arts students who have been misguided in the selection of secondary education subjects, who may do well and be better suited to science and in particular biochemistry than a higher-education arts program. Many students change academic and vocational plans late in their secondary education and higher education, and studies have shown that a significant number of students do not know what they want to do by the time they get to college. Must these students who have not had much science in secondary education be excluded from science in higher education and in particular be excluded from biochemistry? Pearlmutter and Pearlmutter (1977) report in an evaluation of a biochemistry program that students with no prior biochemistry background performed quite satisfactorily in their program. There might be some value in looking at intellectual aptitude as a selection criterion to biochemistry courses for those enthusiastic students who unfortunately do not have conventional requirements for eligibility. Inductive and deductive thinking, numeracy and spatial reasoning are examples of intellectual parameters which impress as being of potential diagnostic value. American college entrance tests, British A and O levels, and similar higher-education academic screening devices do have elements of aptitude in them. These tests of course, are very valuable selection instruments for higher education, but what is being suggested here is that other tests which are more particularly diagnostic of intellectual functioning pertinent to biochemistry should be investigated and developed with the intent of screening the late entrants to this science. Studies of the relationship between intellectual aptitude and biochemistry may also provide some insight into the modes of thinking which underlie biochemistry learning, and which may be useful to curriculum planning and the design of educational materials.

If the use of intellectual assessment does not appeal, perhaps specific tests of basic knowledge and skills pertinent to biochemistry could be constructed to test suitability for entrance into particular biochemistry courses. Construction of such tests is not easy. One would need to hypothesise areas of knowledge and skills which would possibly be fundamental precursors to biochemistry, develop test questions, and then validate the test or tests with students. Such measuring instruments may be useful not only in assessing the suitability of an individual candidate who does not have the conventional requirements for a particular biochemistry course but also may be useful in assessing the overall strengths and weaknesses of a class of students before instruction begins.

As higher-education courses become more tailored to fit individual students needs and become more specifically designed to educate special vocational or interest groups, so the structure of current selection criteria must be examined and continually validated. Are national and standardised secondary-school achievement tests accurate and valid predictors for first courses in biochemistry? Are supplementary assessments needed to augment these general criteria, and to act as screening devices for those students who have not proceeded through education in the conventional way or made career-decisions rather late in their educational careers?

Employment Selection. Not only is assessment used in decisions about entering higher education and biochemistry courses, but it is also an important part of the other end of formal education, namely guidance and selection for employment. Assessment for this purpose is usually cumulative, being collected over the academic career of the student and presented on a transcript or academic record. The quality of grades on the transcript influences job prospects, although the quality of passes is not the sole criterion. Some employers now pay particular attention to personality and the life-style of the student while he was taking part in higher education, such that part-time students are preferred sometimes to full-time students with better grades. The justification for using higher-education academic examination results, and/or personality tests, and/or interview data about life-style for job selection is again a more rational than empirically-based activity. The literature is again replete in predictive studies which empirically examine student success in biochemistry courses and success as a biochemist or other professional for which the course is reputedly oriented. What is the relationship between a biochemistry course intended for nurses and success of nurses in roles which require biochemistry knowledge and skills? We don't really know. Farnsworth (1979) makes the case for appropriate biochemistry education for graduates in industry, and touches on the issue of appropriate biochemistry education for particular jobs.

Those who wish to investigate the relationship between academic success and jobs encounter the problem of obtaining good statistics, similar to the situation described in the studies of the selection process to higher education. Further impediment to predicting job success from examinations is the problem of isolating fair predictive criteria. How does one measure the good chemist, the good doctor, or the good dentist against which higher-education biochemistry results may be compared? Such predictive criterion measures must be free from bias, such that different employment conditions do not influence performance measures unequally; must be relevant; must be reliable, such that the measures of success on the job are stable, reproducible and precise; and must be available and convenient to acquire. Such criteria are difficult to obtain. Criticisms of the inappropriateness of a particular biochemistry course for a particular job are often difficult to substantiate or refute with any degree of empiricism because poor biochemistry course measurement and poor measurement of job performance prevail. How appropriate are our biochemistry examination results for job selection? How well do they predict job success? These are important questions so long as examination results are used in employment screening. The whole area requires careful scrutiny.

Guidance During Biochemistry Education. Not only is assessment part of admission to biochemistry courses and part of selection to employment but it is also important for guiding students while they are in school. Assessment helps in deciding whether biochemistry should be pursued at more advanced levels. It may help a student decide to abandon the subject. These are important personal decisions. Where examination data are used to help make such important personal decisions, particular care is needed in interpreting results, because there is the potential for error in all mental and academic tests. There are basically two different sources of error which will be discussed in this chapter. There are those errors which emanate from the student and those which result from inconsistencies within or between examiners. Examiner error will be discussed in detail later. Student error may arise due to basic idiosyncracies of the student. There are diurnal variations demonstrated by students in examinations, such that the results of a test given in the morning may be different from the results of the same test given to the same student in the afternoon. Extra-curricula events, such as a family dispute or an accident the night or morning before an examination, may have a detrimental effect on the student at examination time. Teachers are aware of hour-to-hour and day-to-day events which affect the concentration and effort of students. Chance, fate, luck or whatever also plays its part. By chance a lazy student may find the examination questions match the limited area he studied, while another student who studied hard may be unfortunate to have prepared for everything other than that which was in the exam. On the objective examination one student may be so shaken by the first three questions that it makes him too anxious and upset for the rest of the test. Another student may find the first three questions easy and so be lifted psychologically for the rest of the test. These are all examples of student error in assessment. Combining the test scores of a number of students theoretically results in the elimination of idiosyncratic error. At any one examination setting, those students who score uncharacteristically high are balanced by those students who score uncharacteristically low. While group scores have this non-systematic error counterbalanced, individual scores have not. Interpretation of an individual's true progress on the basis of one biochemistry examination or even two is too speculative. The way to reduce the possibility of student error of course is to make repeated assessments. By averaging these assessments one reduces the chance of this idiosyncratic error, if it is present, from unduly influencing a decision. It can be seen from the above that there is a danger of a biochemistry course relying upon one final examination as an indicator of a student's true achievement in a course of study.

Sometimes the difference score, or change score between a pre-course test and a post-course test is used by student and teacher to assess suitability of progress in biochemistry. The difference score supposedly reflects the learning which went on in the interim period. Such attribution however, may be a mistake. Student A may score uncharacteristically high on the pre-test and low on the post-test. The difference, or his change score, would be small compared to student B who may score uncharacteristically low on the pre-test and high on the post-test. Individual decisions for either student dictated solely on the basis of these change scores would be misguidance.

Besides the danger of idiosyncratic error affecting change scores, one should be cognisant that different rates of progress may be expected at different levels of academic difficulty. It is usually easier for a student to improve from 30 to 35 per cent correct on an examination than improve from 90 to 95 per cent. For this reason some sort of logarithmic scale, such as that used in the athletic decathlon scoring procedure, would seem desirable where progress comparisons are thought necessary between students who were at different beginning levels. On the other hand it could be argued that it is smart students who are more likely to be at a higher level at the beginning of the academic race, and because smart students learn quicker it would not be necessary to compensate for the difficulty of material which occurs at the higher levels, and so logarithmic scales would be unnecessary before one could make student comparisons. The difficulty for the biochemistry teacher in developing valid tests with associated fair-score scales to compare progress rates of different students for awards or whatever, far outweighs its usefulness. All in all the use of different scores, change scores or progress scores, calculated on the difference between a single pre-course test result and a single post-course test result is to be discouraged for individual student guidance purposes. Progress comparisons between students on the basis of such scores should not be made either. Besides, some wily student may score low on the pre-test on purpose. In principal the greater the number of tests given to any one student without interfering with the progress of learning and without incurring fatigue, the more likely assessment will reveal the student's true biochemistry achievement.

One must be continually vigilant to limit the abuse of tests in the placement and selection of students. The foregoing presents examples of some of the problems in establishing the validity of selection tests. Assessment which is intended to help guide students must not be allowed to develop into assessment for the purposes of assigning students especially when the alternative choices are not equal and the outcomes are irreversible. The consequences of misplacement may be immediate and personal, and the student's self-concept may be badly damaged. There are serious long-term outcomes when misplacement leads to a relatively irreversible career-choice not appropriate to an individual. However, decisions have to be taken on information available. It is best that such decisions, where possible, are taken by the student himself, and that the teacher acting as adviser points out the strengths and limitations of the tests used in the decision-making. The point is not to discourage the use of tests for selection and guidance, it is to encourage the responsible use of them.

Concerning Assessment to Help Guide the Teacher

Are Tests too Widely Used? Some critics of assessment argue that tests are too widely used and dominate the whole academic enterprise. These observations are appealing and without doubt have elements of truth about them. But education is no longer a stroll with Socrates in the Lyceum. The case is more likely one of inappropriate testing procedures and the misapplication of examination results than too much biochemistry assessment. Learning requires questioning, answering and confirming responses. Questioning, answering and confirming describes the testing process too. We memorise best when we review, rehearse or recall, and higher learning processes require evaluative activity. But this sort of testing in learning is personal and informal. A valid criticism of assessment in biochemistry may be that it has become too formal and impersonal and not linked closely enough to learning. Testing certainly becomes a burden when it continually interrupts and interferes with classroom learning. The challenge is to make assessment interesting, fun, personal, meaningful, relevant and a natural part of instruction. The techniques of formal assessment and the social pressures for assessment must not be allowed to dominate.

Assessment Directs Learning. It is claimed that some biochemistry teachers teach only so that students may pass tests, and many biochemistry students learn only when it is necessary to pass tests. There is possibly some truth in both of these allegations. No matter what, these claims draw attention to the powerful influence of assessment in directing instruction and directing learning. To be valid, academic achievement tests must measure the objectives of the instruction. But not only should objectives dictate assessment, the converse is also desirable — assessment should affect the formulation of objectives. If assessment is considered or even planned at the same time as biochemistry objectives are set then some potential objectives which cannot be properly assessed may be discarded from the curriculum or relegated in priority. Some biochemists believe, like Magar (1962), that objectives which cannot be properly assessed have little place in education. This position may be seen as an indictment of the dominating effect of assessment on education. On the other hand objectives which are vague and not clearly defined will remain obscure to students. Is it fair to evaluate students on objectives which are vague and ill-defined? Chapter 1 has discussed in detail the setting of objectives in biochemistry courses. Consideration of assessment when objectives are being set has a very important effect on the formulation of the objectives by helping to define them and by helping to prioritise them.

Students learn best when they are clear about the objectives and direction of their study. Assessment at the beginning of or early in the biochemistry course helps orientate the student to the way in which he will be tested during the remainder of that course. This early testing can influence study and learning style (Frunder, 1978). If early assessment is primarily in essay form the student may learn by writing essays, by discussion, and by formulating arguments. If the assessment is in an objective format such as multiple-choice or true/false, then student learning may take the form of recognition of equations and facts or of verifying the correctness of statements. But should students know beforehand how they are to be assessed so that they may prepare accordingly? Some teachers argue that if students are not forewarned as to the assessment format they will need to prepare for all eventualities. This sort of argument is certainly appropriate when one views assessment and learning as an adverse situation. The argument also has merit if it is thought that students would indeed benefit from being able to use knowledge in a variety of test formats, and where it is possible to test in all of these formats. But often these arguments are really excuses used by teachers who have not bothered or who do not have the time to formulate assessment procedures until late in the course. If it is important that students be able to demonstrate their learning in any of a variety of test formats then ideally they should be told and tested accordingly.

Table of Specifications. An important feature of a good assessment system is that it is balanced such that the questions which are asked cover or reflect the objectives of the course. Important objectives should be given preference or weighting over the less important objectives. A good way of examining an academic test for its balance (content validity) is to draw up a table of specifications. An example of a table of specifications is presented in Figure 5.1. The rows refer to content areas while the columns refer to behavioural processes. Any box in the table is the combination of an expected student behaviour, e.g. knowing facts, and a content area, e.g. proteins. Therefore each box is the rudiment of a behavioural objective. The test builder(s) place in each box a number out of five (five being equivalent to a rating of most important and one being a rating of least important), which reflects the relative importance of the objective designated by that box. When columns are added comparisons of behavioural processes can be made. When rows are added content can be compared. Having set up this ideal model the next step is to build a test on these specifications, or alternatively analyse the test already constructed to see how it fits. The latter is achieved by writing in each box the actual points assigned to the equivalent question or questions in the test. Results can be very revealing and often demonstrate how far our actual test deviates from our ideal test based on the prioritised objectives of our instruction. Sometimes it is difficult to write more than one question to match a particular objective. Balance is achieved in these cases by manipulating the number of points assigned to that question so that it reflects the table of specifications recommendations. Bloom's Taxonomy of Educational Objectives (1956) in the cognitive domain is a source of ideas for the process parameters of the table of specifications.

Different Test Formats. There are different intellectual skills underlying different test formats. Where objective questions such as multiple-choice and true/false require a high degree of comprehension, essay

Figure 5.1: A Model of a Table of Specifications

CONTENT

		Carbohydrates	Fats	Proteins	Sub-totals
B P E R H O A C V E I S O S U R A L	Recall				
	Quantify				
	Analyse				
	Infer				
	Sub-totals				

questions require a high degree of verbal expression. It has been noted by some biochemistry teachers that some students have particular difficulty with MCQs. A common explanation of the problem is that students tend to dwell on the alternatives and become confused with the myriad of less important exceptions suggested by each alternative. Counselling may include convincing the student to approach the test in a methodical way such that only a short fixed time is spent on any one alternative and any one question. The student should be encouraged to select the best answer and not assume that any one of the alternatives is correct in all situations. There are a number of short books which serve the purpose of helping students prepare for a variety of tests. How much of the improvement in test-taking is due to confidence and how much is due to technique is debatable. It is also possible that a better performance following practice may be due to an increase in biochemistry acumen acquired through constant practice in biochemistry tests.

It has been postulated that convergent thinkers are better at multiple-choice questions than divergent thinkers. Whereas the former tend to narrow down on a solution by disregarding the exceptions, divergent thinkers tend to associate freely such that each alternative triggers off a host of possible associated factors. The result for the divergent thinker is total confusion which feeds on itself as he gets further into the test. The divergent/convergent-thinking explanation appears to have merit. It has been suggested that the first few items in the multiple-choice question quiz should be the easy ones or practice ones to get the candidate confident and underway. Can students be trained to think convergently in one situation, such as when answering a multiple-choice question and divergently in another, such as when answering more discursive essay questions? Research in this area may be useful, not just in helping biochemistry students in examinations but also providing some insight into the cognitive processes which underly the way in which biochemistry students learn biochemistry and solve biochemistry problems.

Discussion of the best format to test is frequent amongst biochemistry teachers. The following presents the advantages and disadvantages of objective and essay formats.

Comparison between Objective and Essay Questions. Compared to essay questions, objective questions:

(1) are more quickly and easily scored;
(2) provide more opportunity for sampling a wider content area, thereby increasing validity;
(3) have a higher reliability;
(4) do not favour the verbally facile and those with good handwriting;
(5) are objectively scored;
(6) allow students to score themselves and so have immediate feedback of answers;
(7) demand more specific and active information;
(8) facilitate frequent testing and testing prior to instruction;
(9) may be item analysed;
(10) can be scored by a machine or by clerical help, freeing the teacher for other duties.

Compared to essay questions, objective questions have the following disadvantages:

(1) are more difficult and time-consuming to prepare;
(2) offer difficulties for the student with comprehension problems;
(3) emphasise recognition rather than recall;
(4) may favour the student experienced in test-talking;
(5) may foster poor study habits by exercising trivia and details over principles.

Compared to objective questions, essay questions have the following advantages:

(1) are more easily prepared;
(2) do not necessitate secretarial help;
(3) do not need to be machine-produced, can be presented on the blackboard;
(4) provide an opportunity to measure students' ability to organise, interpret, compare and cite examples;
(5) provide a measure of language knowledge and facility;
(6) may provide less opportunity for guessing;
(7) may give more evidence of depth of understanding of the subject;
(8) offer a chance to describe relevant application.

Compared to objective questions, essay questions have the following disadvantages:

(1) scoring tends to be subjective and takes longer;
(2) have lower reliability, scores are less precise and consistent;
(3) may allow one to sample only a narrow content area, hence reducing validity and reliability;
(4) may be difficult to read and so invoke a negative attitude in the marker;
(5) questions are often not sufficiently specific and encourage glittering generalities;
(6) provide greater opportunity for the halo effect to operate.

Having reviewed the pros and cons of objective and essay formats it can be seen that the choice between these methods of testing is really a compromise amongst alternatives.

Skill-testing. A principle basic to the valid selection or development of an assessment procedure is that the examination replicates as closely as possible the way in which the learning will be used in the real world. For example, if the teaching of a calorific value of various sorts of foodstuffs is important to the job of the dietician, then it would seem important that these figures be memorised and recallable. Multiple-choice or recognition-type activities would not be appropriate in this case as would a fill-in-the-blank type. Some very innovative pen and paper problem-solving tests are reported in the biochemistry literature (Carrington, 1978; Baggott and Trojak, 1978; Miller *et al.*, 1979). However, the selection of testing formats is usually more than just a choice between objective and essay-written types. Skill-testing, especially in the biochemistry laboratory, has always had considerable relevance. If it is important that a biochemist be able to analyse physically the carboxyhaemoglobin levels of blood, then this objective would be better assessed by the student doing the job in a laboratory rather than describing the process in an essay. It is a challenging but worthwhile exercise for biochemistry teachers to envisage the way or ways in which a particular piece of information or skill would be used in the real world and test students in that way or in a way which closely resembles it. Application of this principle is a keynote to valid assessment.

Because it is usually difficult to test in a real situation, simulated testing has rapidly developed as an alternative. It is not uncommon to find programmed patients used in assessing advisory or counselling skills in a simulated clinic (Harden, 1979). Evaluation using good simulated patients is often better than randomly selecting a patient from a clinic because the latter does not guarantee a standardised problem for each student. A big disadvantage of simulation however, is that the examinee usually knows that the situation is a pretence. Success in this artificial environment may be a reflection of the student's acting ability. In some situations, such as teaching and assessing students' ability to handle radioactive materials, simulation may be the only legal alternative. Simulation-testing is governed by cost and convenience. Its availablility as a testing technique is often contingent upon its use and justification as a teaching device as well.

Laboratory, clinical and skill assessment are frequently beset by queries about their reliability more than by challenges to their validity. Reliability refers to the consistency of test scores from one measurement to the next. In skill assessment there are three sources of unreliability. One source is inconsistency of the examiner (within-rater unreliability), a second source of error is inconsistency between two markers (between-rater unreliability), and the third source of error is student unreliability. Student unreliability and idiosyncratic error was discussed earlier in the chapter. Within-rater reliability may be determined by comparing an original score for a piece of work with a second rating of the same work having controlled memory cues in the re-test. Contradictions in scores often occur when performance criteria have not been clearly detailed, and when the instructor becomes fatigued or bored when overloaded with marking. One tends to find this source of error in essay marking, the marking of written project-work, the grading of art objects and aesthetic activities, and judging affective qualities. Error is reduced by practice, by detailing a scoring check-list, by keeping the marking environment free from unnecessary stress and resting when tired, and by checking one's marking at discrete points in the marking schedule. There are a number of mathematical ways of calculating reliability. The simplest way in this situation is to mark the work then change the order and remark a number of coded pieces of the same work, and compute the correlation coefficient for these two lists of scores. There are many education-evaluation books which describe how to do this.

The second source of reliability, between-rater reliability, refers to the degree of consistency between two or more examiners in assessing the same piece of work. An example of error found in this situation as it applies to laboratory testing is where two examiners observe the same student's experiment and allocate different marks for it. A number of studies have looked at this in biochemistry education (Strang, 1977; Nimmo and Flynn, 1978). Differences between examiners can be dramatically reduced if some or all of the following steps are taken. Markers should discuss and agree on the awarding of points before the exam. This includes both listing mark-earning features as well as assigning weightings to each feature, for some part of the experiment may be worth more marks than another. (Similarly, some argument put forward in an essay may be worth more marks than other arguments.) The exercise of examiners marking the same piece of work and discussing differences should promote consistency between them, although Strang (1977) found this not to be so. He found inconsistencies between his colleagues' markings which were not removed as a result of using marking schemes. Because the students who wrote the essays were very high quality and very similar, lack of variability may have been a fundamental reason for the poor correlation found. Besides improving reliability, an additional benefit of examiners getting together to approve the consistency of their marking is the opportunity it provides to review the objectives of instruction and to analyse in detail the structure of some of the tasks being taught.

This section on the use of assessment by teachers has really been a smorgasbörd of a variety of aspects of assessing. Much of the discussion pertains to other purposes of testing besides that of guiding the teacher.

Concerning Assessment in Innovation and Experimentation

'Formative' and 'summative' evaluation are two jargon terms found in educational assessment literature. Formative evaluation essentially refers to that ongoing, sometimes informal, assessment carried out while a project is progressing, while summative evaluation relates to more formal assessment much of which comes when the project is finished. It acts to summarise the effects of the program. Distinguishing between these two types of evaluation is useful when planning assessment for innovation and experimentation purposes in biochemistry education. Techniques of formative evaluation include short tests, questionnaires, structured group discussions and interviews (sometimes tape-recorded) between teachers, students and resource people involved in the experiment. Often measures of affect and attitude are employed to help assess the comfortableness, appropriateness, perceived usefulness and acceptability of parts of the new program as it develops so that immediate adjustment may be made. New education programs seldom work satisfactorily the first time. Initial difficulties are occasionally so obvious that a complex formative evaluation system appears unnecessary to decide whether to continue or abandon some projects. But it is in the realms of finer tuning of experimental programs that formative evaluation can be very useful. Summative evaluation is a more structured, more objective and more formal assessment procedure. Data may be reported in terms of cognitive and psychomotor performance changes, feelings about the program, personal and administrative factors and economic parameters. Formative evaluation is usually expected with all educational innovation. Pearlmutter and Pearlmutter (1977) used a variety of assessment procedures to measure a self-study biochemistry course for medical students. They used pre-tests, student ratings of teacher effectiveness, national standardised tests, home-made MCQs, and questionnaires at three- and one-week periods. The evaluation associated with some experimental programs can be quite extensive.

Common to both formative and summative evaluation is the problem of the assessment procedure contaminating the learning exercise when the assessment is not meant to be part of it. Every assessment is potentially a new learning, relearning or review activity. It is not uncommon to find in the results of the supposedly controlled educational experiment a dramatic improvement in the scores of the control group, who of course did not receive the experimental biochemistry program. The improvement occurs not because of student maturation or because the control group has taken alternative educational programs, though this must be watched, but because pre-experiment evaluation (pre-testing) arouses the curiosity of some of the students in the control group with the effect that these students make efforts to satisfy their curiosity and independently seek verification of their pre-test answers. When the pre-test is later repeated as post-test some of the control group score much better as a result of their independent efforts. This phenomenon explains the lack of difference found between controls and experimental groups in some educational research. The consequence is potentially serious for it may give rise to a decision not to adopt what really is a better new innovation. The experimental problem can be alleviated somewhat by developing 'different but similar' pre- and post-tests, instead of using the same test before and after the education, or by including a non-pretested control group in the experimental design.

Pre-testing may also contaminate the scores of the experimental group. In an earlier discussion it was noted that tests are used by students as cues about how to study, and used to indicate the objectives of the course. Knowing the objectives of the course affects the learning process, for we learn better when we know where we are going. For this reason pre-experimental assessment cannot be divorced from the experimental education and should be kept in the new program when it moves from trial to permanent status. If the pre-test is not desired in the final program then a more complex experimental assessment needs to be developed. Pre-test contamination of later learning has its counterpart in other formative assessment activities too. Mid-course assessment may act as a review exercise or have a motivating effect on both students and staff for subsequent learning. If mid-course assessment is not intended in the final program then the trial program must be re-evaluated without it.

Another distressing feature of education experimentation in biochemistry as well as in other areas of education is the unavailability of data about the long-term consequences. Many teachers and administrators appear to be satisfied with the short-term outcomes. There are problems in interpreting data when assessments are made one month, three months, six months and twelve months after the end of the experiment. As mentioned previously, an assessment may be a new learning, re-learning or review situation, so that each time the evaluation is made on the same students it acts as a refresher activity and affects the results obtained in subsequent assessments. For example, assessment at three months helps review and thereby affects assessment at six months. To avoid such dilemmas cross-sectional assessment at different post-program intervals using different students who have been through the same experimental program would be desirable. The practical logistics of organising such a protocol may be one of the reasons why the long-term effects of biochemistry education is seldom documented.

An important part of assessing innovation has to do with affect. Did the students enjoy the method? Did they find it boring? What parts did they find a challenge? What parts did they feel were too easy? Cognitive or academic test results reveal some of the problem areas. Affective assessment may also help to reveal additional problems and advantages, and suggest why some of these good or bad spots occur. Affective assessment may take the form of discussion and interviews with students and staff. Where good relationships prevail such a direct method is often all that is needed. But where those involved don't know each other or where there are too many students or staff to interview in the available time, then pencil and paper questionnaires or more formal affective tests may be useful. If there is too much distrust and suspicion by students or staff, which sometimes arises when considerations are made of future outcomes of an experimental program, then indirect assessment of affective qualities is unlikely to be valid. Some people answer not only in ways which they think the examiner wants to hear, but often in ways contingent upon their perceptions of the consequences of the experiment. Where distrust and suspicion prevail direct measurement is usually better than indirect methods.

Participant observation using neutral observers has been taken from anthropology and the social sciences and used as a technique to avoid some of the problems associated with getting honest answers to questions of feeling. Participant observation is the technical name given to the procedure where one or two assessors would take the bio-chemistry program with the students, observing and recording their own and students' reactions as they proceed. But of course participant observation is not a pure measure either, for assessor bias is always there. Observers bring their own likes, dislikes and perceptions into every assessment they make. People see what they want to see. It is recommended that if participant observation techniques are to be used, then incorporated on the panel should be a number of people who are from different backgrounds, with different ideologies and who do not have any real stake in the outcome of the experiment. This recommendation certainly applies to the large, expensive and important assessments. But in most biochemistry classroom experiments a friendly fellow-teacher or two may serve a useful function in the role of participant observer(s). It is helpful for the observer to have some check-list to aid his analysis — some predetermined list of factors which the experimenter wants the observer to pay particular attention to. Check-lists may be used at the end or during the experimental trial, though care must be taken in the latter case not to distract students.

What are some of the attitudes and values that may be related to learning and teaching biochemistry? Cataloguing potential non-cognitive parameters and tests which may have relevance to bio-biochemistry education is beyond the scope of this book. However, O. K. Buros' Mental Measurement Year Books are potentially useful reference sources. The reader is also directed to psychologists and educationalists as a source of guidance, for the study of human behaviour does not have the equivalent of chemistry's periodic table of elements. Many psychological concepts are operationally defined and terminology is not clearly standardised in meaning. Krathwohl and co-workers (1964) produced a taxonomy of educational objectives in the affective domain similar to Bloom's taxonomy in the cognitive domain. The affective taxonomy suggests a conceptual framework to review affect. It points out some of the difficulties in definitions but is very useful for those who wish to get a foothold in the area. The taxonomy, though intended for measurement purposes, has been criticised on the grounds that it does not reflect real psychological entities. For example it does not help to assess quality, degrees, or strength of feeling about an issue, but tends to be more lateral, expansive and broad in its definition of affect. This criticism may be valid but it does not detract from the book's value as a source of ideas presented in a clear, uncomplicated fashion. Rezler (1973) also provides a simple and clear exposition of methods of assessing attitudes as they apply to health and medical education. This article outlines fundamental properties of different scales and questionnaires such as the lykert scale, semantic differentials, closed-form questionnaires, observational rating scales and check-lists.

Even though there may be some uncertainty about aspects of the measurement of affect, biochemistry teachers should not be fearful of the area. If a commonsense approach is taken, simple questionnaires and check-lists can be constructed which give some insight into student receptivity for new ideas and established practices. It should be clearly noted that assessments of attitudes and values are not appropriate criteria upon which to award student grades. However, affective assessment is valid and appropriate if one is assessing instructional goals for programs and not for individual students. The care that is taken in the selection and construction of tests of affect to measure the effects of programs depends upon the importance of these measurements in deciding whether to abandon or adopt the innovation. Where decisions are important experts should be called in.

Some 'Newer' Assessment Procedures. Another way of interpreting innovation in biochemistry education assessment is innovation as it pertains to assessment methods. With new ways of teaching bio-chemistry and new teaching materials it has become possible to achieve a close and more natural integration of assessment with learning. An example of this integration is in the presentation of instruction in a modular form, where the course of study is divided into self-contained small units of instruction with self-assessment at the end of each module. The modules usually are arranged in a recommended learning sequence. This form of curriculum organisation is often difficult to administer but provides greater flexibility to the students in learning, and the manner of assessment allows one to check the quality of this learning as one progresses. The assessment procedure is sometimes called continuous assessment and it may contribute substantially to the final grade of a formal biochemistry course.

After a series of modules a comprehensive assessment may be provided to achieve integration and act as a review mechanism. A major problem of assessment with a self-instructional modular course is the opportunity provided for students to cheat. The occasion may be seized upon when passing examinations is seen as the most important aspect of study. The problem is partly resolved by developing item banks (Buckley-Sharp and Harris, 1970; Morgan, 1979a; Bryce, 1979a; Haywood and Wood, 1977), or a series of identical tests so that students do not know exactly what will be asked. But developing banks of objective items is a lot of work, and it is usually not long before students have all the questions anyhow. Some academics argue that memorising many questions is worthwhile in its own right, for in learning the questions and answers one really is acquiring what was intended by the instruction. The dilemma of invalidating assessment when students 'rob the bank' of objectives questions may be resolved somewhat if assessment goes beyond the objective format such that short-answer essay and skill-testing methods are used. When non-objective questions are used the examiner must be readily available to mark work otherwise self-pacing and student control of his own learning is lost.

The computer has been used in assessment experiments in higher education with varying degrees of success. Some are reported in this book. The PLATO system is an example of a well-developed, higher-education, computer-assisted learning system used in the USA. Continuous objective assessment is a basic feature of this system. The word-processor appears to have a definite future in expanding the range of assessment activities for computer-assisted learning enterprises. The computer has accelerated the possibilities of tailoring tests to meet individual student's needs. In the testing situation the student progresses through a test ordered according to difficulty till he has difficulties. When the student gets a question wrong a similar question of the same difficulty level is asked. If the student gets the answer correct he moves on to a more difficult question. If he fails to get sufficient correct answers at any one stage of the testing the computer recommends that he takes some remedial learning. The computer may even tell the student where this remedial material can be located. This sort of mechanised testing has potential for biochemistry education in the future, but it is expensive to develop. Such procedures would require widespread use to be justified, which infers that there would need to be greater standardisation of biochemistry instruction content. This may not appeal in an educational climate of individualised instruction for special professions and purposes.

The future of higher education may involve more emphasis being placed on continuing education. One particular form which this may take is called distance learning. It has been widely discussed in medical education circles in Britain and in the education of professions allied to medicine. Though distance learning is not likely entirely to replace conferences, it may become a more convenient and financially acceptable form. There are assessment considerations for distance learning which relate to recertification. The philosophy, psychology and validity of recertification assessment will not be pursued here, but assessment as it pertains to distance learning of a voluntary nature will. It is generally agreed that higher-level learning involves the development of creative and original ideas, and that many important decisions are compromises and value judgements. Harden *et al.* (1979) have gone about distance learning in a novel way in continuing medical education. They have developed a sophisticated 'readers digest' approach. Case studies and questions on a particular theme, such as care of the diabetic or diagnosis of the alcoholic, are sent to doctors. They answer the objective questions and send their responses to a compilation centre where the answers are averaged. Concensus answers are then reported back to the doctor. He compares his original answer with the consensus of which his original response forms part. The method may be criticised in that it has the potential to homogenise the medical procedure, but doctors are free to accept the answers and their own answers are identifiable only by themselves. Distance learning of this form seems to have a lot of potential especially where expense has to be contained and where certification or recertification is not required.

In this section some of the implications of experimental design in the appraisal of educational experiments and innovation has been made. The reader is encouraged to apply those same strict criteria used in biochemistry research in the appraisal of educational experiments. He is encouraged to seek help from education and behavioural scientists in regard to those aspects of educational research that he is not sure about. It is hoped the foregoing also gives some idea of the variety of innovation in assessment in biochemistry education and the types of problems being accounted.

Program Evaluation

Accountability is part of higher education, since some assurance of the usefulness of state-sponsored educational programs seems reasonable. For example, assessment may provide information to senior teachers, administrators, government officials and the like about whether a new biochemistry program or an established curriculum is working well. Evidence for this purpose may be obtained by listening to the testimonials of teachers and students. But such a method is notoriously biased. One way to supplement the information gained from testimonials is by systematic and regular testing of students using valid and reliable biochemistry achievement tests. The strengths and limitations of these tests have been mentioned earlier. Sometimes national standardised academic performance tests are used. The assumption underlying the use of these tests is that the good biochemistry teacher will produce good student performance and vice versa. While this is generally true, one must be careful in interpreting the results because so many factors affect the quality of student learning. The quality of teachers, the availability of laboratory equipment and facilities, the attitude and ability of the students in the class, and more important the different objectives from biochemistry class to class will all affect assessment outcome. When one is using test performance to assess any one of these factors the results are being confounded by the others. Where national standardised academic tests may be used to compare institutions they do little to reveal those things that contribute to the differences. Take, for example, the use of the student performance on national standardised biochemistry achievement tests to appraise the quality of two different biochemistry teachers who do not teach in the same institution. Colleges, because of their geography and location, differ in the quality of students who apply and enter their classes. We know high-ability students tend to learn faster than low-ability students. Comparing two teachers in institutions not comparable in student ability is not fair. Iborra and Lozano (1980) compared biochemistry in two Spanish medical schools. One school was superior to the other, but the assessment did not reveal why such differences should exist. One of the criticisms advanced against national standardised tests in biochemistry and other areas as are found in the American Medical National Boards, is that tests tend overly to influence the objectives of biochemistry with the effect of it homogenising biochemistry curricula nationally. This abuse of tests is possible where ignorance of testing prevails. Comparisons of colleges on the basis of these tests, however, are sometimes useful as motivators.

It is possible to train students in achievement test-taking. Hyde *et al.* (1976) demonstrated this when they trained second-year medical students over a two-week period to take multiple-choice questions in Part 1 of the American National Boards. The national ranking of the institution changed dramatically for the better as a result of this training exercise. Care must be exercised not to allow national standardised tests to dominate completely our biochemistry courses.

Student Rating of Teacher Effectiveness. There has been interest in recent years in the use of student ratings of teacher effectiveness as part of program evaluation and teacher guidance. The basic intent of these ratings is to find out what the students think of the teacher and the course. In some colleges ratings are compulsory. Assessment usually consists of a series of statements which reflects some dimension of the teaching. Usually a five-point scale is used. Sometimes a free-comment section is included at the end. Costin *et al.* (1972) list some potential uses of student ratings of college teaching in their review. They suggest that these ratings can be used by the teacher for his own benefit and by the administration to compare individual teachers and departments. They see the inventories being used to provide information on relative strengths and weaknesses, and perhaps being useful to students in selecting courses where options are available. The article also provides a source list for a variety of instruments used in rating activities. The general concensus about such rating scales are that they are very useful. In their summary and conclusions, Costin *et al.* say:

> research findings suggest that the criteria used by students in their ratings of instructors had much more to do with the quality of the presentation of the material than with the entertainment value of the course *per se*. Such attributes as preparedness, clarity, and stimulation of students' intellectual curiosity were typically mentioned by students in describing their best instructors.

Correlations between course rating and grade received, when observed at all, tended to be small, and several studies suggested that such correlations resulted in greater interest in the course by the students receiving better grades, rather than from a 'reward effect'.

It would seem undesirable to use ratings scales of teachers' effectiveness as the sole criteria for promotion of teachers however. Other factors which should be taken into account in promotion include participation in other non-classroom teaching activities, requests to give guest lectures, development of new teaching materials and methods, and other extra-curricular activities including research.

Appraising Education Materials. As inventories have developed for the rating of the effectiveness of the teacher, so inventories have been developed to help in the assessment of educational materials. Films and tape/slides are some of the methodologies appraised in this way. There are difficulties in getting concensus among teachers on the acceptability of many audio-visual aids. There is often some small trivial part of the content which puts a reviewer completely against the program. However, instructional material rating forms may be useful in screening out the extremes. Ideally the best assessment of educational materials is by using preview and student performance data. Where pre- or post-test control groups are used in this evaluation there is sometimes difficulty in getting enough questions from one film or tape/slide program to make a precise measurement. Thus the problem is that the existence of no difference between the pre- and post-test or between an experimental and control group may be because the measuring instrument is too crude to pick up the finer differences which really exist.

A large and important part of program evaluation which will not be discussed in this chapter concerns cost-benefit and cost-effectiveness. Economic analysis is becoming increasingly important in higher education. With an increase in such analysis has come a realisation of the fundamental aspect of assessment in higher education, namely that there will always have to be the unavoidable problem of having to make value judgements about the data collected. One must clearly differentiate the process of collecting data from the process of decision-making on the basis of that data. Where we should seek to be as objective as we can in the former it is logic and value judgements which dominate the latter.

6   SIMULATION AND GAMING IN BIOCHEMISTRY

Michael Tribe

> But though learning may be conferred by solitude, its
> application must be attained by general converse. He has
> learned to no purpose that is not able to teach; and he will
> always teach unsuccessfully, who cannot recommend his
> sentiments by his diction or address.
>
> Samuel Johnson (From Essays 'The Adventurer'
Saturday, 19 January 1754)

Introduction

Simulation and gaming may at first appear very strange activities to be
associated with the teaching and learning of biochemistry, a subject
with a high factual content and a strong practical base. However, in
learning about the content and principles of any subject, especially one
with such wide application as biochemistry, it is all too easy to
concentrate on small items in great depth to the exclusion of the
broader perspective to which these items relate; put another way,
students often see each item as a separate entity rather than part of a
coherent whole. This is not to say that the 'in-depth' approach to
topics within the subject should be abandoned, but rather that the
interrelationships and broader perspectives should not be ignored. The
carefully-considered use of simulations and games in undergraduate, and
indeed postgraduate, biochemistry courses, is one way of remedying
this problem. Furthermore, because much of our teaching is directed
towards the individual student, whether this be as one of a cohort in
lectures or practicals, or as an individual concerned with a research
project, simulation games offer by contrast an opportunity for group
learning. It is certainly true that the development of independence in
the learner is of paramount importance. But more recently, with career
prospects in mind and the requirement to work as part of a group or
team, it has been argued that teaching should not only be directed
towards developing independence in the learner, but also towards a
degree of interdependence (Postlethwait *et al.*, 1969; Becher *et al.*,
1973; Tribe and Peacock, 1973). The balance between independence

and interdependence in student learning should be a fundamental part of our educational thinking, for as Postlethwait *et al.* (1969) have pointed out: 'Educational research has demonstrated that no single medium can produce the broad range of responses necessary to achieve mastery of a complex subject . . .' Simulation and gaming can assist in achieving this balance.

Definitions

Since the terms 'simulation' and 'game' sometimes lead to confusion, it is perhaps appropriate at this point to provide a definition of each term. The definitions that follow are those of Gibbs (1975):

A *game* is an activity which is carried out by co-operating or competing decision-makers seeking to achieve their objectives within a framework of rules. In contrast a *simulation* is a dynamic representation which uses substitute components and relationships to replace their real or hypothetical counterparts.

In many instances, there are elements of both simulations and games, so that the term 'simulation-game' is an aptly descriptive one (Bloomer, 1973).

Background to Simulation and Gaming

Simulations are extensively used in training programs, particularly those involving fast response and co-ordination of motor skills to potentially hazardous situations, such as piloting an aircraft or driving a motor car. On the same principle that medical treatment is equally hazardous in the hands of the novice, clinical-problem simulations have been developed to teach biochemistry to medical students (Jepson and Smith, 1972, 1973; Blanchaer, 1975; Vella and Martin, 1976; Winkler, 1978; see also Vella in Chapter 4 of this book).

With more widespread availability of small-size computers (and even programmable pocket calculators!), many biological scientists have been encouraged to develop 'model' situations within areas of biochemistry, genetics, physiology, population ecology and evolution (Crosby, 1961; Appleton, 1973; Tomlinson, 1976; Summers and Summers, 1976, 1977; Hull, 1977, 1978; Bryce, 1977; Hancock, 1978; Cunningham,

1979; Kidd, 1979; Morgan, 1979a; Shone, 1979). Since the use of computers in biochemistry teaching is discussed in Chapter 9 of this book, the subject will not be pursued further here.

Games are much older in origin than simulations, although the distinction between the two is not always clear-cut. They were originally used to develop skills in military strategy, but within the past two decades have entered the general educational arena, first into business and management studies, then into some aspects of the secondary-school curriculum, and eventually into higher education. Simulation-games have been most extensively used in areas such as applied economics, political strategy, land planning, social administration and medical diagnosis, where students can practise role-play and decision-making without threat to society or the environment (Armstrong and Taylor, 1970, 1971; Smith and Jepson, 1972; Zuckermann and Horn, 1973; Horn, 1977; Taylor and Walford, 1978). In the pure sciences, particularly subjects in which there is a high factual content, games and simulation have at best been viewed with caution, at worst with suspicion and hostility. It seems important therefore to outline those features about the structure of simulation-games which merit their use with students who are learning about science.

In 1973, Tribe and Peacock suggested that simulation-games provide the following:

(1) structured situations which encourage scientific discovery and problem-solving;
(2) motivation and stimulation for students to learn about a subject in which they had previously little or no knowledge;
(3) motivation for learning factual information when it is related to a specific problem or task;
(4) an effective way of learning how to communicate and negotiate with other people in a team or group, and the enjoyment that follows;
(5) insight into the social aspects of learning, particularly by encouraging co-operative learning that leads to the establishment of working and social friendships which carry over into 'unscheduled' hours;
(6) decision-making opportunities, normally without censure or threat;
(7) a closer empathy between students and tutors;
(8) a means of altering attitudes of students who have participated in the simulation-game.

In the view of the author, these attributes of simulation-games still hold true, although the list reveals several items that could be (perhaps often are) achieved by other learning situations. Nevertheless, the carefully-constructed simulation-game reinforces these features within a unique decision-making context (Dowdeswell, 1974; Atthill, 1975), thus providing students with a much better understanding, appreciation and perspective of the subject.

The Design and Objectives of Scientific Simulation-games

Characteristics

The first thing that one can say about simulation-games is that they possess a number of critical characteristics, which can easily be remembered because they all begin with the letter '*C*'.

The first is that all simulation-games involve some element of *conflict*. Such conflict may relate to the persons playing the game, or the resources, or the time limits, or any other external conditions imposed by the game. In many biochemical simulation-games the conflict is often one of deciding whether the 'research team' should co-operate or compete with others, or alternatively the conflict is one of priorities when time and resources to complete the task are limited. Secondly, simulation-games have to have rules, and these rules impose a *constraint* upon the ways in which the participants behave. Thirdly, it is essential that some of the basic elements of a simulation-game *correspond* with their 'real-life' counterparts, although in other respects the situation may be artificial or *contrived*. The degree of contrivance or correspondence can vary considerably.

Finally, a simulation-game must have a point of *closure* and usually a means of determining 'winners' and 'losers'. The end of a simulation-game may be decided by a time limit, or when specific objectives have been achieved, or when all but one player or group has been eliminated. In many cases, 'winners' and 'losers' are easily determined. In other cases, where complex, open-ended problems are to be solved, a concensus view may have to be sought as to the person or group that has made greatest progress in arriving at the best solution to the problem.

Structure and Design

In terms of structure, scientific simulation-games normally take the following form, irrespective of the problem posed and the roles played by students. The terminology used here is adapted from Boocock and Schild (1968).

1. Introduction: (i) to the problem; (ii) the roles to be played; (iii) decision on the number of players or the size of groups; (iv) location of resources.

2. Execution: (i) the course of action to be taken on the problem; (ii) decisions regarding the organisation of each group; (iii) communication of ideas and actions; (iv) whether to co-operate or compete with other individuals or groups.

3. Analysis: (i) evaluation of collected observations and data; (ii) presentation or reports, conclusions or 'original' papers; (iii) inter-group discussion.

4. Follow-up: (i) comparison with real-life situation if possible; (ii) supplementary reading.

These four headings provide a useful guide to both the participants of a game and to teachers embarking upon designing a simulation-game for the first time. For the teacher, it enables the 'gaming' portion to be separated from the 'content' or 'simulation' portion (Horn, 1977). Consequently, attention can be focused on any behavioural activity that the teacher wishes to encourage or develop in the students and then the teacher can decide what kind of game situation is best suited for the task. Horn (1977) summarises these ideas very succinctly in terms of the *behavioural focus* and the *vehicles of attainment*, i.e. 'frameworks for learning that use the powerful incentive of small group co-operation and competition in an orderly fashion'. In all events appropriate 'content' can be added later.

To illustrate this point, if the behavioural focus is simply collecting, sorting and learning scientific facts, then a simple competitive card game might be the appropriate vehicle. For example, Carrington (1978) has outlined a number of simple undergraduate card games to assist in the learning of biochemical formulae and their names: 'The object is to increase the exposure of biologically important compounds to students' (in mind not body!). In complete contrast, the 'Ecology Game' and the 'Enzyme Game' developed at Sussex University (Tribe and Peacock, 1976; Peacock and Tribe, 1979), have as their behavioural focus three major objectives:

(1) to provide opportunities for learning effective participation in task-orientated groups;

(2) to enable students to explore a real scientific problem in both breadth and depth;

(3) to provide opportunities for using scientific concepts already familiar to them in a broader context.

The vehicle for attaining this behavioural focus is the 'research team', operating under the constraints of time and available 'research funds'.

Objectives

The importance of objectives appropriate to academic games has been emphasised by Cowan (1975). The high motivation and enjoyment of participating in games is important but not sufficient. It is also essential that the game has academic merit and that students actively learn from it. For this reason, Cowan (1975) has suggested that successful academic games should have the following features:

(1) Properly defined objectives;

(2) A likelihood that the objectives will be achieved;

(3) Sustained motivation ensured for a sufficient learning period;

(4) A relatively low requirement for initial skills and abilities;

(5) Scope to expand the difficulty of the game as the players improve;

(6) The facility for remedial or supplementary teaching if necessary;

(7) Proper feedback to the teacher;

(8) Realistic, natural and relevant game activities;

(9) Constant interaction between the players;

(10) Simple rules, equipment and scoring.

Cowan concludes that of the two games that he has developed for undergraduates in civil engineering, one of them 'LOAD' (see Cowan, 1975) fulfils most of those objectives desirable in a successful academic game. The two simulation-games developed at Sussex (Tribe and Peacock, 1976; Peacock and Tribe, 1979) — the 'Ecology Game' and the 'Enzyme Game' — also meet most of the criteria listed above, although the 'Enzyme Game' requires that students have at least a basic understanding of enzyme structure and function before they start. Obviously, most students can get a great deal more from a game if, from the point of view of content, they begin with a sound understanding of the concepts and principles underlying the problem. This is particularly true for simulation-games of the open-ended type (see Gibbs, 1975), because it allows them to be played at a more sophisticated level. Indeed, one of the reasons for designing research simulation-games at Sussex, was to enable students to apply the basic knowledge that they had acquired from self-learning texts to decision-making in the game situation (Tribe *et al.*, 1975). This background is not an obligatory pre-requisite for the 'Ecology Game', but it does provide both useful revision of previously-learnt ideas, as well as putting these ideas into a new context. These games are 'won' therefore by *skill* and not by chance.

Although pre-knowledge is often of considerable advantage, the *follow-up* to the game in the form of laboratory work, extended reading, or ecological field trips is of even greater value, in that many of the game's objectives are further reinforced and can be successfully recalled by students two or three years later. Some outline will now be given of some simulation-games relevant to biochemistry.

A. The Enzyme Game (Peacock and Tribe, 1979)

Students are divided into groups, each group comprising about six members. Each group is then presented with a *Students' Guide* to the game, a *synopsis of useful biochemical techniques* (in which are summarised 24 techniques of importance in enzymology) and a handout of *previously-published information* on the crude enzyme extract.

Within the *Students' Guide*, the problem, roles to be played, objectives, and structure of the game are outlined as described below.

'The Problem'

An enzyme has recently been discovered in bovine pancreas which exhibits proteolytic activity. As yet, no purification scheme exists for this enzyme, although it can easily be assayed spectrophotometrically using synthetic peptide or ester substrates. In the game you are to consider yourself part of a research team beginning an investigation into the structure and mechanism of action of this enzyme.

The way in which the game itself is organised is described below. However, this game is a dynamic situation in which the tutors' role is to provide the research groups with the resources they need to fulfil the objectives, without giving direct help to them. The tutors may therefore modify the structure of the game as situations develop, either on their own initiative or in response to proposals from research groups.

Objectives

These are to determine as far as possible the structure and mechanism of action of the enzyme.

Methods of Investigation

At the start of the game each group will be provided with a booklet giving a synopsis of useful techniques for studying the enzyme and a card summarising previously-published information on the enzyme. You must then decide what information is required to fulfil the objectives of the game. Experimental information can be obtained at the expense of part of your group's research grant (details below). As the tutors have a data bank consisting of over 50 cards, it is important to *specify in detail* the experimental information required, otherwise it is impossible for the tutor to select the most relevant data card. All data collected should be kept in the folder provided to prevent it from being mislaid.

Time Limits

The game will be divided into sessions of approximately 1.5 hours, each representing one-year's research time. The tutors will tell you when the sessions are to be held and how many there are to be.

Research Grants

These are given by the 'Research Council', which is represented by the tutors. Each group is initially given a grant of 20-weeks' research time (symbolised by coloured cards) by the Research Council. You are expected to have spent most or all of this by the end of the first session. Further funds for the next session are allocated on the basis of the Annual Report submitted by each group at the end of the session (see later). 'Publications' in the *'Journal of Enzymology'* automatically increase a group's research grant for the following session (see later).

Reports

Annual Reports. Before the end of each session (by a deadline set by the tutors), each group must submit in writing an Annual Report. This should summarise the group's past progress and future plans. It should not exceed one side of A4 or American quarto paper.

Final Report. At the end of the game there will be a final conference of all the groups and the Research Council. Each group is expected to have prepared for this a verbal summary of its success in meeting the objectives of the game. It should not take longer than approximately five minutes to deliver.

Publications

A research group can submit, at any time, a paper for 'publication' in the '*Journal of Enzymology*'. If accepted, the paper is then available for other groups to read and it automatically increases the publishing-group's research grant for the next session. A publication must be original and should summarise the interpretation of specified data cards. It should not be longer than one side of A4 or American quarto paper. The tutors will act as journal editors in deciding the quality, originality and suitability of the paper for publication.

The previously-published information on the enzyme, which is given to the students at this stage, states the following:

Previously-published Information on the Enzyme

The enzyme can be isolated from bovine pancreas where it occurs as an inactive proenzyme. The proenzyme can be activated by incubation with 1% trypsin (a proteolytic enzyme).

The active enzyme hydrolyses the two substrates, carbobenzoxy-glycyl-L-tryptophan and hippuryl-glycyl-DL-phenyllactate as shown in Figure 6.1.

In addition to the Students' Guide, there is also a Tutors' Guide, which gives more detailed information about the problem and advice on how to administer the game. The 'Enzyme Game' has been extensively tried out at Sussex University and at Leicester University by my collaborator Dr Derek Peacock. More recently the game has been undertaken in other institutions of higher education, quite independently of its authors. For those interested in the views of these independent trials, see Bryce (1979e); Dalziel (1979); Gayford (1979); and McBrien (1980).

B. The Ecology Game (Tribe and Peacock, 1976)

Whilst the 'Ecology Game' (by definition) is not directly concerned with biochemistry, there are aspects of the game which have their roots in both biochemistry and environmental chemistry. For this reason, it has been briefly included here, particularly for biochemists contemplating the use of simulation-games, who wish to see the scope of open-

Figure 6.1: **Substrates Hydrolysed by the Proteolytic Enzyme Under Study in the Enzyme Game (a) a Synthetic Tripeptide (Carbobenzoxy-glycyl-L-tryptophan) and (b) an Ester (Hippuryl-glycyl-DL-phenyllactate). The Arrow Indicates the Bond Hydrolysed by the Enzyme**

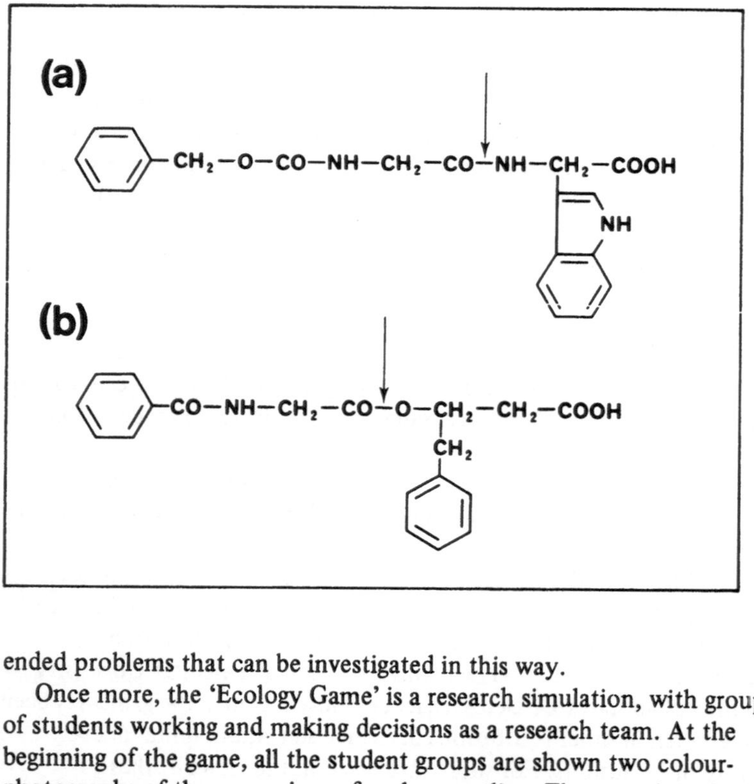

ended problems that can be investigated in this way.

Once more, the 'Ecology Game' is a research simulation, with groups of students working and making decisions as a research team. At the beginning of the game, all the student groups are shown two colour-photographs of the same piece of rocky coastline. They are then given a map of the area and a Student's Guide containing the following extract:

'The Problem'

The problem is a very real one, which actually occurred in Britain. The two photographs show the same piece of coastline taken in July 1960 and in July 1967. There are obvious differences between the two. You are asked to play the role of the ecologist who has visited this coastline fairly frequently and who would be interested in trying to find out the differences between the two situations and the factors which caused them.

Objectives

These are:

(1) to determine what changes have occurred in the ecosystem;
(2) to investigate possible reasons for the changes;
(3) to recommend suitable courses of action in a similar situation.

Instructions on the methods of investigation, time limits, research grants and reports are very similar to those already given for the 'Enzyme Game', and because of the earlier publication of the 'Ecology Game', it has received very extensive trials, particularly at Sussex and at the Open University summer schools in Britain.

C. Research Games: A Complement to Laboratory Project Work

Details have been given of two research simulation-games developed at Sussex University, but the concept of research games is certainly not unique to us. Suckling *et al.* (1979) at Edinburgh, Hultquist at Michigan and Vella and Martin (1976) at Saskatchewan have used somewhat shorter research games to achieve many similar objectives to the ones that have been outlined in earlier sections of this chapter. In a very similar way to the 'Enzyme Game' (Section A above), the aim is to provide perspectives on a problem and to condense to a matter of hours the experimental time which in real life would take a year or more:

> the student does not perform any laboratory experiments but otherwise carries out all the normal steps in a research project, planning experiments, interpreting the results and drawing conclusions before the design of the next experiment. The results are provided by a lecturer based on a model of the system being studied which he has devised.

A selection of projects (research games) used by Suckling and his colleagues includes:

(1) enzyme regulation in a bacterial metabolic pathway;
(2) the biochemical basis of the polymorphism in the concentration of erythrocyte glutathione observed in sheep;
(3) the effect of cytochalasin B on red-cell carbohydrate transport.

D. New Ideas in Simulation and Gaming for Biochemistry Students

In contrast to the two published simulation-games mentioned in A and B above, other unpublished ideas are worth including here because of their different behavioural focus.

One idea developed by Dr Derek Peacock and his colleagues in the Medical Biochemistry Department at Leicester is to put groups of third-year students in the position of a committee whose role is to consider research grant proposals. Although all the applicants and their places of work are fictitious, the research proposals and costings are all quite credible. Students have to discuss and decide, giving their reasons, which of the research proposals they are willing to support within the total budget that has been allocated to them by the Research Council. The decisions reached by the student groups are then compared and discussed in conjunction with the biochemistry staff.

Another game that we have pursued with first-year students at Sussex is based upon the idea of a 'Treasure Hunt'. 'Clues' are given to small groups of students (two or three students per group) in the form of journal or book references, catalogue numbers, microfiche references, etc. From each successive reference, students have to collect salient points from summaries, or key items of data from graphs or tables in order to be able to build up a report, which is submitted to the tutor for marking and comment. The area that we have examined in some depth through this approach has been 'The Use of Biological Organisms as Sources of Energy', because information is scattered in a large number of books and journals. The main objective of the exercise is to familiarise first-year undergraduates with the university library and the methods of cataloguing and retrieving scientific information — a most important aspect of the educational development of any individual.

Merits of Simulation and Gaming

Developmental testing of simulation-games of the kind described above with different groups of students for nearly a decade, confirms that a number of skills are developed, which, whilst not unique to research simulation-games, are much enhanced by them. For example:

(1) the ability to formulate precise questions appropriate to the problem under investigation;

(2) to formulate hypotheses and design appropriate experiments to test them;

(3) to recognise the significance of each experiment and its implication to the whole problem;

(4) to interpret experimental data accurately and thoroughly;

(5) to recognise the cumulative power of learning;

(6) to write short, concise reports and conclusions on experimental findings;

(7) to communicate ideas and listen to reasoned opinions within a group;

(8) to be able to summarise clearly and accurately the collective experimental findings of a group and defend them when challenged;

(9) to give confidence in tackling complex problems.

The fact that these skills are developed in a stimulating and enjoyable way (as indicated by comments from student questionnaires), argues strongly in favour of the careful use of research simulation-games in biochemical education.

Although many tutors contemplating the use of games and simulations are often hesitant and even resistant 'to the concept of "gaming" as a pedagogic activity in higher education' (Taylor and Carter, 1971), confidence in designing and handling games is greatly assisted by attendance at workshops. Indeed, many converts to simulation and gaming originally attended workshops from hostile curiosity!

From experience of using research simulations, the advantages to tutors are that: first they can be used most effectively with students of mixed abilities and backgrounds; secondly they can be adapted to different levels of attainment by making the rules and conditions easier or harder; thirdly they allow students to apply different skills and abilities in a group decision-making situation; and fourthly they allow new experimental data and information to be added when they become available (Tribe and Peacock, 1973).

Administrative Problems

Running simulation-games properly within a fairly conventional teaching timetable of lectures and practicals can present problems. With games of relatively short duration (one hour), there are fewer difficulties if tutorial periods are an integral part of the timetable. For exercises such as the 'Enzyme Game' however, there is the difficulty of finding available time in an already overcrowded schedule, because with such games, two- to three-hour periods on consecutive days may be desirable. If the game forms an integral part of a timetabled course, provision and planning can usually be made in advance. Failing this, either a block of evening tutorials (as at the Open University Summer Schools) or a weekend residential course could provide suitable alternatives.

A second problem is finding a suitable location for a simulation-game, where competition (or co-operation) between groups are involved. Ideally, separate but adjacent rooms around a central resource area are advocated. A third problem relates to the number of student groups that a tutor can effectively handle. The experienced tutor will probably manage three groups at a time, i.e. a staff:student ratio of 1:18. For less experienced tutors, it is suggested that no more than two groups per tutor (i.e. a ratio of 1:12) be employed for smooth running of the game.

Criticisms

Some people might object to biochemical games, because such games are not the same as actually working in a biochemical research laboratory. This comment is true, but the objection is based on a misunderstanding of the game's objectives. As Boocock and Schild (1968) point out:

> the object of the game is to involve the student in the types of situation, motives, practical constraints and decisions that are the subject of study, not the specific details. The student should emerge from the game with a better understanding of what it was all about, what was possible and what was not, and why.

The fact that simulation-games are novel, motivating and enjoyed by most students makes many academics suspicious of their value in higher education; even the word 'game' conjures up a picture of levity and entertainment. Cynical observers also point out that despite intense involvement by student groups with the problem, decisions may not be responsibly taken, since 'the risks and dangers are no more than "paper" consequences' (quoted in Taylor and Carter, 1971). There is of course, no reason why learning experiences should not be enjoyable.

Indeed, research suggests that if 'something which students enjoy doing can be introduced relevantly into learning, the resistance [to learning] is reduced and efficiency is likely to be maximised' (Beard, 1970). On the second point, it is sufficient to say that reckless decisions in simulation-games would soon lead the group into 'disaster' as gauged by tutor feedback, so that realisation of the risks and dangers, even in terms of 'paper' consequences, are better than no realisation at all. Nevertheless, any novel form of teaching and learning carried out by enthusiasts, is likely to meet with success — the so-called 'Hawthorne effect'. The true test of whether games can make a significant contribution to biochemical education is for less committed teachers to *try them* and if necessary modify them to their local needs and conditions.

A further criticism is that considerable preparation time and effort are involved for only a small component of the science course. However, if there is a serious deficiency in the scientific education of our undergraduates that can be remedied by the use of simulation-games, then the time and effort is well spent.

Evaluation of Simulation-games

Evaluating the effectiveness of simulation-games, both in terms of behaviour and content, is not an easy task. Not only is it difficult in some instances to measure whether the game's objectives have been achieved, but also and more importantly, whether they could have been better achieved by alternative teaching strategies.

It clearly befits advocates of simulation-games to demonstrate that these are worthwhile components of a biochemistry course, or indeed of any course in higher education. Therefore a quantitative evaluation of the elements of games is likely to be more persuasive to those contemplating their use.

On the other hand, as Cowan (1975) points out, there is a need to be wary that we do not just:

study the behaviour of the [game's] participants, and then review and discuss these details with almost clinical detachment, expressing [our] views in a complex psychological jargon. Surely the learning of the participants should be the priority, unless the game is intended merely as an educational experience for the observers.

In attempting an evaluation, consideration should be given to three things: (1) evaluation of the scientific content of the game; (2) evaluation of the behavioural objectives of the game, including the group dynamics and structure of the game; and (3) evaluation of simulation-games as a teaching technique. At times, it is not easy to distinguish clearly between these three components.

Regarding the first of these and to some extent the second, there are several methods that can be employed. In the 'Enzyme Game' for example, 'annual reports', 'original papers' and 'final reports' (as outlined in the Students' Guide) can provide collective evidence, whilst a student post-test or examination can provide information about individuals. Although there is an element of subjective assessment by the tutor regarding papers and reports, most of the data collected in this way correlates well with the data obtained from individual students. More important, however, is that comment and criticism by the tutor on reports and papers provide immediate feedback to the students as the game progresses.

With respect to group dynamics and structure of the game, at least two methods can yield quantitative data: (1) closed-circuit television and video-tape monitoring, and (2) the student/tutor assessment schedule. The author has not had experience of the first method, but interesting work in this area has been carried out by Dimitriou (1971) and others (see Taylor and Walford, 1978). We have, however, experimented with the second method. For example, *group dynamics* can be broken down into more specific items, with each item rated on an ascending five-point scale as shown:

Group dynamics	Scale				
	v. poor ⟶ v. good				
(1) Listening	−2	−1	0	+1	+2
(2) Communicating ideas	−2	−1	0	+1	+2
(3) Understanding before action	−2	−1	0	+1	+2
(4) Decision-making	−2	−1	0	+1	+2
(5) Conflict-handling	−2	−1	0	+1	+2
(6) Division of labour based on strengths within the group	−2	−1	0	+1	+2
(7) Organisation of group's ideas	−2	−1	0	+1	+2
(8) Self-conscious evaluation of performance by group	−2	−1	0	+1	+2

(If one or more of these items is/are considered to be more important than the others, then larger values can be given to each point on the scale, i.e. −4, −2, 0, +2, +4 etc.)

After the game is over, individual students are asked to rate each item subjectively, based on their own experiences and impressions of working with their peer-group. Tutors involved in the game also make a subjective rating of the way in which groups appeared to organise their activities, the order in which 'research' was carried out and the quality of 'reports' and 'papers' submitted. Collectively, this information can be used to modify and improve the game. It is therefore a positive (though not definitive) approach to evaluating many of the important aspects of simulation-games.

Regarding the third aspect — evaluation of simulation-games as a teaching technique — such evaluation or comparison is difficult and lacks precision of measurement, a point made by several authors (Garvey, 1971; Rackham, 1970). Indeed, there is little empirical evidence to indicate whether simulation is more effective or less effective than other teaching techniques (Garvey, 1971).

Conclusions

From the statements made above, it should be clear that there has been no attempt to prove that simulation and gaming are better than other forms of teaching and learning, even if this were possible. Secondly, simulation-games must not be regarded as a gimmick, but an alternative means of presenting problems and ideas, for as Tansey and Unwin (1969) report:

the whole of simulation is merely a means to an end. It is an alternative strategy. If the lecture . . . or any other method works there is no need for simulation. If on the other hand, these methods are not achieving the results that are desired, then it is as well to have an alternative method of presentation. Simulation is merely this method of presentation.

Since other forms of learning rarely make provision for decision-making, there may be an increasing demand for the availability of suitable simulation-games. The situation is well summed-up by Beard (1970):

Simulation systems and games, like programmed learning are methods tailored to the specific needs of particular groups or students; but, unlike programmed learning, they provide for the achievement of objectives which are otherwise impossible or difficult to attain in practice — such as 'decision-making' in dangerous circumstances — and so fill gaps in a number of educational programmes. They tend to share the disadvantage that preparation time is consuming; but, once prepared, they usually prove more absorbingly interesting and enjoyable than programmes. Where programmes on the whole are intended for individual use, simulation systems are used by groups and help to educate students in handling group interactions. Finally, they share with programmes the advantage of being carefully designed and so are well suited to experimental techniques, they can, therefore, be more easily assessed and improved for the attainment of particular purposes. With so many advantages, it seems reasonable to expect a proliferation of these techniques . . . within the near future.

Acknowledgements

I am greatly indebted to my colleague, Dr Derek Peacock for his friendship and collaboration.

7 LEARNING IN SMALL GROUPS

Alec E. Wood

Introduction

Most biological subjects have traditionally been taught as practical
subjects with a lecture component (or the reverse). More recently there
has been a tendency to introduce a tutorial into this program to help
achieve some particular objectives. This has resulted in a greater
flexibility of the teaching schedule and some questioning of the aims
and objectives of each part of the program. The purpose of this chapter
is to examine the role of tutorials in the context of the teaching of a
whole subject. Because of some confusion of terms, the discussion will
primarily be concerned with small-group learning, the most common
form of which has long been described as the tutorial. As this book
aims to be more practically orientated and because of my own background, the discussion will be mainly practical rather than theoretical
and will emphasise approaches which have been used in some areas of
biology or which are obviously usable in this discipline.

Aims of Lectures, Practical Classes and Tutorials

In teaching a subject one usually hopes that students will gain certain
information about a discipline, acquire certain experimental skills,
learn how to acquire and interpret practical and experimental data and
learn to understand and generalise about the given area of knowledge.
From these general guidelines, it is useful to see what part is played by
the various methods of teaching.

Normally, laboratory work is the main means by which students
gain technical experience of a whole range of procedures and
equipment. This also gives experience in obtaining and recording results
with subsequent interpretation. Usually it takes quite a large proportion of the time allotted to the subject, but this is seen to be
worthwhile because of the skills learned and the experience gained in
obtaining and interpreting experimental results. Sometimes there is
also an opportunity to plan experiments. Some of these procedures can,
of course, be carried out by small groups. It is possible for an

experiment to be planned and carried out by three or four students and this can be a very valuable exercise in group activity, communication of ideas, shared responsibility and corporate planning. It is also possible for some group activity to be used at the end of the practical period so that results can be assessed and discussed and possibly further experiments designed and full implications explored. This can be very useful to ensure that laboratory work does have a definite set of conclusions.

Lectures are mainly regarded as the primary means of conveying information and it is usually hoped that they will also promote thought and understanding, though the evidence for this is not encouraging. As a means of providing information, their effectiveness obviously varies with the lecturer, the general conditions, etc. The lecturer may simply use spoken presentation or any one of a range of visual methods for conveying information (up to and including colour video segments). Very often the presentation is largely or completely a one-way process and may continue undiminished for the full 50 minutes (irrespective of the maximum concentration span of students!). Various procedures have been tried to increase feedback in lectures. The effectiveness depends on the particular techniques, the size of the class, the physical surroundings and the style of the lecturer. There are some cases where particular lecturers have been able to develop effective group participation in lectures and this area is one that could be explored and developed with profit. The whole area of lectures is well discussed by Bligh (1972), who includes a whole series of suggestions for obtaining feedback.

One of the pressing problems with lecturers is the matter of defining their objectives in relation to all the other information sources. These sources range from textbooks through reference lists and annotated reading lists to the various kinds of programmed texts and integrated texts, with visual material either as slides, cards or video-cassettes. Faced with all these self-paced alternatives for information input, the lecturer needs to assess all the methods open to him and then define the objectives of his lectures and the material which will be covered by other methods. Some background to self-paced learning systems, including the Keller plan, can be found in Boud *et al.* (1975).

Tutorials have been mostly used as a group project with a tutor, to solve some particular task which may be either general or quite specific. This whole area will be dealt with in more detail. Much of the relevant literature can be found in the bibliographies edited by Powell (1971, 1977).

Aims of Small Groups

When small-group work is scheduled within a course, the objectives of such groups need to be set out clearly. In some circumstances groups may be used to answer questions and problems that have arisen in lectures. Sometimes they are used to solve a set of problems. For this, the students work through the problems individually and then check their answers with a tutor and/or hold a general discussion at the end of the period. More often, the groups are used to solve some kind of task, by means of group discussion, with a tutor present, with the view being taken that this is a more effective method of promoting understanding.

The whole area of aims is well discussed by Beard *et al.* (1978) as part of a more general treatment and there is also much helpful material in UTMU (1972) which is a set of papers based on the proceedings of a conference. For valuable insights into all aspects of group teaching, the works of Abercrombie (1979) and the general discussion by Beard (1976) should also be consulted. A general survey is provided by Ottaway (1968) which is a good introduction. He points out that group discussion has special value because of the opportunity for full participation by students in a free discussion. Here, students have equal rights, are expected to learn from each other and give their own judgement.

In most cases where small groups are used in teaching biology, the group centres on some task, while the role of the tutor is variable (see later). While the task is the focus, another aim is to help students to gain understanding through group discussion and in the process to develop critical thinking ability and the capacity to integrate ideas and draw deductions and conclusions. Hence there are the long-term objectives of developing mental skills and the more immediate ones of solving a specific task.

Functioning of Groups

The literature on group functioning is immense and only a brief sample will be indicated. Some of this is much more theoretical than the rest, but some concept of how groups function is almost essential if the best use is to be made of group learning. A particularly useful earlier book is that of Hill (1962) which highlights critical features for good group functioning and outlines a set of procedures which should help a group to function effectively. There is a useful analysis of specific and overall roles within a group, with some non-functional ones also being highlighted. While the discussion is not extensive, the general outline is extremely useful. Another good general background to group work is provided by Klein (1961) where a discussion of the categories of group interaction helps clarify the objectives of a group. There is also a useful examination of the social context of the group, in terms of leadership, structure, function and morale. The book by Argyle (1967) is less related to learning through small groups, but it does provide much theoretical background for understanding group functioning. There is a good discussion of social techniques, social behaviour and social skills. A fascinating chapter deals with eye-contact and includes a description of the normal pattern with a discussion of the motivational basis for looking. This provides a valuable introduction to an important aspect of interpersonal contact within groups.

The whole area of encounter groups contributes valuably to a consideration of group learning. A useful introduction is provided by Rogers (1970) who describes how such groups may be facilitated and how people may change as a result of their experience in such groups. While the whole approach is centred on intensive group experience rather than on content or group activities, there are still many useful insights into group processes and procedures. Helpful insights could be transferred to groups with quite a different context and structure. In another work, Rogers (1969) questions many of the assumptions of current academic procedures and argues that the most valuable learning is that where students are personally involved, are self-initiating, and where they are provided with a choice of options so that they may develop their own goals and select their own pattern of learning. This approach has many ramifications and a specific philosophical basis, but it does raise important questions. Most courses are fairly rigidly structured, but it is not clear that students could not be offered much more choice as to what they learned and how they went about it. Rogers argues persuasively that such a program would lead to a greater student response and be a more meaningful learning process.

The work by Bramley (1979) contains a fascinating description of a series of tutorials where the author acted as an observer in tutorial groups and later discussed the session with the tutor. There is a useful description of group structure and how over a period of time it was modified and also how the role of the tutor changed. Some of the conclusions are that group functioning is aided by: the development of receptivity or passivity; seeing the group as a system rather than solely as a set of individuals; welcoming emotionalism. This latter may seem strange in science-based groups, but if it is taken in the sense of expecting a response and an involvement in the group processes of the whole person and not just a mind, then this could be a valuable balance to dull intellectualism and remind us that participants in groups are real people. Other parts of the work deal more with theoretical concepts and here much of profit can be gained. The concepts of the theme (It), the individual (I) and the group (We) are useful in helping to see the need to keep a balance between the various aspects. There is a discussion of how an understanding of Basic Assumption Groups and Focal Conflict Theory may have valuable insights for group learning. While some of the concepts may need modification and simplification, they provide useful methods of interpretation of certain group activities. There is also a discussion of three modes of group functioning – student to tutor, group to tutor and student to student. The point is made that each mode has its own advantages and may be used for specific purposes and also that a tutor often changes from one mode to another during a single tutorial.

The work of Ruddock (1978) is somewhat unusual, for in addition to presenting problems and identifying issues, there is also the presentation of extensive evidence which provides the basis for the conclusions. There is a good discussion of the role and responsibilities of the leader, with an outline of the options open to him in terms of role – instructor, participant, model, devil's advocate, chairman, consultant. This is a useful section as it focuses attention on the role of the leader in the group and suggests that there should be more clarity in understanding this role. There is also an outline of the responsibilities of the leader in maintaining group functioning. While this assumes that the tutor will function as the leader (which may not always be the case, see below), it does provide a valuable outline of leadership responsibilities.

The work of Abercrombie and Terry (1978) is again based on extensive annotated reports of group discussions, with extensive use being made of video-taped materials. There is a discussion of how content and process may be controlled and group discussion facilitated. There is also a discussion of the first ten minutes of the tutorial and how this is often of crucial importance for the development of the remainder of the discussion.

Much of the theoretical work may seem somewhat remote from real group situations, but the insights gained about the factors which influence group structure are important and can be directly applied. Most of the work described has been carried out within disciplines other than biology, but the same general conditions and constraints operate. The goals and objectives may be weighed differently and there may be more emphasis on content rather than on process, but the principles underlying the effective functioning of the group remain the same.

Types of Small Groups

Problem-solving

In problem-solving groups, a specific problem is set to be solved, either individually, or as a group. Often the whole process is not of a discussion nature, though there may be discussion of the various answers. This approach is well-known in some areas of mathematics, but it can also be applied to certain kinds of tasks in biology. Some interesting work has been reported by Lee (1978) where examples are given of types of group discussion in microbiology. Normally related to specific problems, they also raise general issues in discussion. A good working handbook provided by Ogborn (1977), covers practical aspects and provides considerable depth of treatment. While it is produced by a group of physicists, it is capable of wider application, particularly within science. The goals of groups which function in this way need to be clarified. If the objective is to obtain the correct answer, then it functions satisfactorily. If, on the other hand, the aim is to promote understanding, the obtaining of a correct solution may not indicate that this has been achieved. In addition it may not really be a group activity, so that the positive gains from group interaction may not occur.

Directed Group Discussion

This is probably the most common form of small-group discussion. If students are unused to group work, they need some introduction to the role of the tutor and their own role(s) in the group. It may in fact be very useful to spend some time in introducing the group to group processes and illustrating the needs of the group and how they can be met. The role of the tutor also needs to be outlined and accepted. This can alter during the course of a single tutorial in response to the happenings in the group and it may also change over a period of time as the group matures into a coherent whole and so the tutor is able to modify his role. An extremely interesting account of some of these aspects recorded by Bramley (1979) illustrates how the role of the tutor may change as a result of insights provided by an experienced colleague. In another interesting study recorded by Dee (1976), senior biology students were given a series of group discussions to integrate the diverse strands of their knowledge. Because of the nature of the objectives in this case, no specific preparation was needed and the discussions aimed to help students to make deductions, test assumptions and make connections. The students saw the series as having helped them to think critically and to work out their own ideas. This illustrates that one of the particularly useful roles of group discussion is to help in critical thinking and interpretation, rather than to function as a mechanism for transmission of information.

Non-directed Group Discussion

This kind of group work has numerous forms. In some cases it has been carried out without a tutor being present (Powell, 1974). This work illustrates that such a procedure could be useful and meaningful. Certain aspects have been criticised by Ruddock (1978) who felt that in this approach, the impact of the leader on the group was undervalued. However, it demonstrated that real learning could occur without a tutor and also showed the undesirable effects of domination of the group by the tutor.

Most other forms have a tutor present who functions as a participant rather than a director. Most of these approaches stem from the important work of Abercrombie (1960). Here, the discussion was based on seeing differences, drawing inferences, clarifying the meaning in the case of key words, extrapolating and predicting, and evaluating evidence. All of these concepts are of importance in biochemistry and are central concepts in critical thinking and higher conceptualisation. The argument is presented that group discussion improves judgement because when students are given the same stimulus they produce alternative judgements and these are exposed in discussion. This is important because group discussion centres on the different perceptions which students have of the same subject, whether written or visual. These differing perceptions then have to be modified or defended in group discussion. Although most of Abercrombie's work was concerned with visual stimuli, the same is also true of perception of written tutorial material, though the differences may not be so dramatic.

This approach has been modified and used extensively with first-year biology students — see Robbins (1972), Fleming and Stuckey (1972), Ramsay *et al.* (1978) and Wood (1979). In this work there was some modification in the role of the tutor and specific tutorial material was developed. The tutor was used in the group as a participant rather than as a director or leader. The object was for the tutor to aid the functioning of the group without becoming an authority figure or a source of information. On the other hand the tutor did not become inactive but was able, by suitable questioning and probing, to maintain some direction in the group and also to make some contribution to the level of the discussion. This meant that the various roles of group functioning had to be accepted by the group. Sometimes the leadership role emerged naturally, while at others it was a shared function which was exercised on different occasions by different members. A group can function in this manner only if it can be persuaded to accept this procedure and if the tutor can adopt this role. Often it does mean an apparently slower rate of operation with sometimes the necessity of allowing chaos and error to persist until group members accept responsibility. In this kind of situation, some introduction to group functioning is generally helpful to the group as it enables them to see ways in which they can operate.

With groups who functioned in this way, the type of tutorial material used was very important. One experiment was conducted to study how various types of written material influenced the operation of the group (Fleming and Stuckey, 1972). Four formats were used: general discussion questions; multiple-choice (where factual statements were followed by a set of inferences with the instruction to determine the relationship between the two); experiments, where factual information was given and generalisations had to be drawn from this information; pragmatic, where general principles were given and these had then to be applied to specific problems. It was found that the 'experiments' and 'pragmatic' formats were the most successful. Subsequent experience has indicated that detailed tutorial material is particularly important for this style of tutorial. It must be interesting, pitched at the appropriate level, not too long and preferably become more difficult as one progresses through a topic. Given the availability of such material, it is quite possible for the group to function as an effective learning group, without the tutor directing operations. It helps for the objectives to be clearly specified, so that at the end of the tutorial the students can assess how far these objectives have been achieved. However, the development of group structure requires considerable effort and this coherence is often quite fragile.

Syndicate Groups

This is basically a variation of non-directed group discussion. The basic work, with an evaluation, is provided by Collier (1966, 1969). The essentials of the method are for small groups of students to work through assignments, having been given questions and references. From this discussion a group answer is derived which is subsequently reported back to the larger group and discussed. The tutor is largely independent of the small-group discussions, though he may circulate among the groups. Judging by the responses of students, there was evidence of a greater feeling of involvement and although they seemed to feel that they worked harder, they had also used their books more effectively and were more intellectually stimulated. The method could be used to integrate and compare various aspects of a whole subject, but this would need skill in framing the questions and careful choice of references. Overall it was found that the particular advantage of this approach was that it generated motivation and involvement in the process of active academic study. This method is more suitable for senior students and the task needs to be very carefully chosen, with relevant references. Given these requirements, it is easy to see how this would be a most effective way of involving each member of a small group in the learning process. This approach could be used to solve some biological problems though considerable ingenuity would be needed in devising the assignments.

Discussion of Scientific Papers

An interesting and different approach has been pioneered by Epstein (1970), but it does not seem to have been used as widely as one would have hoped. It sets out to convey to students exactly what a biologist *does* as it is argued that there is a greater need to stimulate curiosity and motivation in students than to revise the informational content of courses. The program was developed for first-year biology students who did not intend to major in biology. More recently it has been tried with other and more senior groups. The approach is for the teacher to select up to ten scientific papers which in some way form a sequence and which are then read and discussed in the group to see how the work was carried out and the logical stages by which it developed. The primary interest is in examining the method of operation of the scientist, not in the facts of science. The first paper is dealt with without any introductory discussion. This approach means that particularly in the early papers, a great amount of time is spent defining terms, but this has much more meaning as it is done in the context of understanding actual scientific work. The sessions are conducted through students' questioning, with no pressure on students to participate or work. There are many digressions which are welcomed, but controlled, so that they do not go beyond ten minutes. Because of this, the first paper may take three to four weeks to complete, but later papers require no more than two weeks. This method places students in contact with real science and arouses their curiousity so that they are prepared for more abstract information.

Later developments included the use of this approach with first-year courses for biology majors, where it was again found to be very effective. In many ways it was easier because of the better background of the students, but the initial rate of progress was slower due to more digressions and the greater desire for information rather than insights. This was controlled by increasing the rate at which the papers were studied. When such students progressed to senior courses, there were some gaps in their informational background, but this was quickly solved because the students had a much greater ability to see what they did not understand and ask very effectively for the necessary clarification. Where the series-of-papers approach is used for senior courses, the papers need to be very recent and the students, because of their greater expertise, tend to become more interested in analysing the experiments rather than in the activities of the experimenter. Some evidence is presented to indicate that once students have one course with this approach, sequel courses are not especially worthwhile, possibly because curiosity and motivation have been stimulated. However, the evidence on this point is not particularly strong and there seems no good reason why sequel courses should not be effective, provided an appropriate set of papers is chosen.

The whole approach would seem to have great potential in experimental sciences such as biochemistry where a sequential series of papers could be devised in many of the major areas of the discipline. This would have the advantages of introducing students to scientific papers at an early stage of their studies, showing them the way scientific discoveries are made and illustrating how one discovery can lead to a whole series of new investigations. While the use of this approach might mean that some of the factual information might not be as completely taught as by more traditional methods, it would seem that this strategy has great potential for involving students in understanding how science works and from this curiosity leading them further into both the process of investigation and the results which have been obtained.

Discussion of Practical Work

Smythe (1974) describes the use of group discussion at the end of practical classes in first-year biology to enable concepts raised and results obtained in the laboratory work to be discussed and clarified. While the period involved was only half an hour, it was felt to be a worthwhile operation. As far as possible the tutor did not participate in the group activity except in a brief summing-up at the end of the discussion period. The design of the discussion topics was found to be critical and needed great care. It would seem that more use could be made of group discussion at the conclusion of practical classes in many experimental subjects as a method of assisting in understanding both the results obtained from experiments and also the theoretical concepts behind such work. It has the advantage of discussing the students' own results, with an increase in motivation and involvement.

Integrated Courses

There have been approaches which have integrated the various parts of courses more completely. The most outstanding work of this type is seen in the Audio-tutorial Method developed by Postlethwait (1969) where lectures and laboratory work have been integrated so that each student works in a booth where audio tapes are played and appropriate equipment is used. This approach has been used in a wide range of subjects, but of particular interest is the report of its use in teaching biochemistry (Garland *et al.*, 1977). Significant differences were found in the response of various course groups. Medical and dental students responded better, possibly because of their more homogeneous background, shorter-term expectations of biochemistry and a strong vocational trend. Science students responded differently, probably because of their more diverse backgrounds and more mixed motivations. The design of an appropriate course of experimental method for science students presented more problems.

A general review of this approach is provided by Dowdeswell (1973) who outlines some of the problems with this method. He particularly questions the effectiveness of this approach for teaching broad concepts, principles and relationships and suggests that often the information gained can easily become a series of semi-independent packages with little connection. He also notes that this method, for a variety of reasons, has been little used in the teaching of higher years.

Other relevant literature is surveyed by Boud *et al.* (1978) where the references relating to personalised systems of instruction and the audio-tutorial method are collected and annotated.

A related approach is reported by Blunt (1976) where a whole anatomy course was integrated around small-group discussion and final sharing in somewhat larger groups. Specific objectives were provided for each session and while there was much informational material to be mastered, there was also much integration and interpretation. This provides a very interesting example of a method of group learning where much information (often derived from visual materials), had to be mastered and this information was then integrated and co-ordinated in small group and plenary sessions.

Another variation is seen in the work of Brewer (1974, 1977) where plant anatomy was taught by means of audio-tutorials, much in the style of Postlethwait. This provided integrated learning of a strongly informational subject. The work also included a weekly small-group session where the week's program was covered informally. The form of the group session was that of student direction with limited tutor assistance and the aim was to provide scope for interpretation and application. Most of the discussion centred on visual materials which were integrated into a quiz which covered the work for the week. The role of the tutorial became more important in later work and more attention was given to the role of the tutor (to promote discussion, as a resource person and to promote group attitudes) and to greater clarification of the objectives of the tutorial. While some of the questions for the tutorial were concerned with recall of knowledge, many of the questions in the later section of the group session required intellectual skills. This work provides an interesting example of how programmed self-instruction and group discussion can be integrated into a single course program. While the subject matter was fairly strongly informationally-based, it should be possible to adapt this approach to more experimental subjects. However, it is not clear whether experimental courses would profit from such highly structured programming. It is evident that great care needs to be taken to try to ensure overall comprehension and integration of material which is covered by any type of modular or mini-course approach.

Practical Considerations

Certain practical constraints obviously affect the functioning of group discussion. Relatively little has been written in this area, though some aspects are raised by Ruddock (1978) and Wood (1979). The size of the group is quite important, for if it is too large the group will tend to fragment into sub-groups, while if it is too small there will not be enough variety to provide the range of inputs necessary for good group functioning. Good comfortable surroundings can also greatly aid group discussion as can the way in which the groups are formed. Some method of formation which uses, at least in part, some of the social groupings which students already have in other contexts can be of assistance, particularly in the early stages of a course of tutorials. Attention to detail in these areas can make the learning much more profitable.

Preparation of Tutorial Materials

Where the tutor does not lead or dominate the group, the written (or visual) tutorial material becomes very important in providing guidance and structure for the discussion. Very little has been written on this aspect and even fewer actual examples have been published. Some general matters are dealt with by Fleming and Stuckey (1972) and Wood (1979) while some examples are provided by Lee (1978) and Smythe (1974). The material needs to have specific objectives, to be written in an interesting way and be of a genuine discussion type rather than simply an information-pooling device.

Experience would tend to suggest that the skills needed for preparation of such material are quite specialised and while the skills may be learned to a fair degree, some people have much more ability in this area than others. It would seem sensible to use staff with these skills very extensively for the preparation of such materials, even when they are not subject specialists. It also suggests that collaborative projects between institutions and publication or circulation of such materials could be of great assistance.

Training of Staff

For many staff, some form of training is helpful if group discussion is to be made more profitable. Staff must first be convinced of the potential of group learning and then they must learn some of the necessary skills. Abercombie (1979) comments that staff must learn how to work in groups, learn to understand the theory behind group functioning and understand the factors which influence groups. Bramley (1979) has a good outline on leadership requirements and also has a useful list of general requirements which many will find useful for guidance in assessing their own functioning. It is also pointed out that the tutor needs accurate self-knowledge as to his own idiosyncrasies which may determine the kind of group he makes. One way of learning to function in groups is to form a group of tutors which will function as a group to achieve this understanding. A report of such a group is provided by Abercrombie and Terry (1978) where areas such as the supportive nature of the group, development of self-awareness and empathy with students are discussed very helpfully. This approach is probably one of the most useful methods as it provides practical experience of group functioning which illustrates many additional features in the process of discussion. While there is a place for specialised input of information regarding both the theoretical background and practical aspects of group functioning, there are advantages in learning much of this by actual group discussion.

For many academics, to become a sharer in a group, rather than a dispenser of knowledge, is a new experience. There is evidence that most can learn to transform their role, though some find it more difficult than others, this being related both to age and personality characteristics (Wood, 1979). However, if staff wish to learn, experienced help is usually available. A critical appraisal of group work by an experienced colleague who sits in on tutorials should prove extremely helpful, even if somewhat traumatic at times.

Evaluating Results

The first task in evaluating group discussion is to decide what is being evaluated. In some cases primary concern will be with process, in other cases with content. If one is primarily concerned with the growth of the group and the individuals in the group, some useful pointers are given by Ruddock (1978) and a list of criteria is provided by Hill (1962). There is still very much to be learned in this area and Abercrombie (1979) wisely remarks that more assistance is needed from social scientists to guide the proper assessment of group functioning. For the detailed analysis of group processes, several procedures are available and an introduction to these is provided in a set of conference papers (UTMU, 1972). These provide a good introduction to Flanders' Interaction Analysis and Bales' Interaction Process Analysis. The former is more appropriate where there is substantial leadership of the group by the tutor while the latter is better suited to the analysis of free discussion.

If the primary interest is in progress of learning of content, this may be evaluated in various ways. The works of Blunt (1976) and Brewer (1977) provide some evidence of better learning through small groups.

However, the testing is complicated. If the evaluation is concerned with a rate of learning of specific information this can be tested and compared fairly readily. However, if the object is to see whether higher learning such as the understanding of principles, integrating knowledge, learning to think critically and drawing deductions, has taken place, this is much more difficult to evaluate objectively. Most often one can only gain general impressions from discussions with students and questionnaires. From this kind of evidence it would seem clear that students can (but not always do) find small-group discussion a valuable means of learning to think constructively and grasp general concepts and principles.

Advantages of Small-group Learning

The use of small groups in learning should be examined in the context of the total objectives in teaching a particular subject and the range of techniques which are available. Valuable guidance may be found in Beard (1976) and Abercrombie (1979) to help obtain a general view of the possibilities. It is not being suggested that small-group teaching should constitute the total learning experience, as other methods may be better suited to the achievement of certain objectives.

The possible use of group discussion in a teaching and learning situation needs to be explored on the basis of the objectives which are desired, and the possible strategies. The objectives should be related to course content and/or the intellectual development of the student. If it is desired to further the development of mental skills such as critical and analytical thinking, to develop communication skills and specific skills such as interpretation and deduction, then the use of group discussion would be most appropriate. If it is desired to encourage understanding through active involvement and by this to increase the motivation of students, then group discussion would be most helpful. This has been written without any reference to the informational content of various subjects, because group discussion can have a valuable part in any discipline. In the sciences there is a need to integrate specific facts and see overall patterns and general principles. In the development of these broader conceptual aspects small-group learning would seem to offer outstanding opportunities. What is needed is the development of suitable materials and skills for the best use of this approach.

8 THE USE OF PRINT IN IMPROVING LEARNING

Alistair M. Stewart

Introduction

Although printed material does not have the glamour of some of the other media it is, nevertheless, one of the most important and readily available instructional media resources.

It can be deduced from Bligh (1972) that printed material is at least as good as lectures when it comes to imparting information. The addition of visual illustration could, conceivably, make it as useful as tape/slide, but it clearly cannot compete with film because it cannot easily portray dynamic processes.

However, if print is to be used successfully, it is necessary to consider: the purpose for which it is to be used; the structuring necessary to facilitate learning; the effect of its legibility and layout on learning; and the role of visual illustrations within the text.

Forms of Print and their Utilisation

There is a tendency to consider the 'book' as the primary printed instructional resource because in every field, no matter how specialised or advanced, books have been written. But there are other forms of printed instructional materials. Carroll (1974) rejects 'anything that can be produced by a printing press' as a definition and chooses, instead, the descriptive term 'frozen language', pointing out that it is stable and enduring and its use is not restricted to particular times and places. Rothkopf (1976) regards print as 'cohesive presentations of some matter which, in style, more or less resembles connected speech' and identifies three different kinds of print: general instructional materials such as textbooks and didactic expository articles; specific instructional works such as programmed instruction; and materials that are the collective experience of a culture, such as literature, and scientific and technical works.

In higher scientific education there are probably three main types of printed instructional materials which students use: textbooks or other expository materials; lecture handouts; and worksheets. *But aren't lecture handouts frequently similar to expository materials? Need they be?*

Perhaps a better classification of printed instructional materials would be a simple division into expository and interactive materials. *But aren't textbooks both expository and interactive?*

A recent book on instructional techniques in higher education (Kozma *et al.*, 1978) states:

> Thus, it can be seen that a book, unlike a lecture, is an active medium, requiring the student's continual engagement in order for the message to be transmitted . . . However, . . . it is not interactive. The potential for the student to alter the message to fit his or her needs is limited to the search or review of some useful subset of the materials that is available. (p. 163)

Hartley (1978), however, does not distinguish between textbooks and interactive materials, arguing that printed 'instructional materials are tools for use in a highly interactive and relatively unpredictable sequence of events' (p. 13).

There is little merit in trying to classify instructional materials in terms of their 'interactiveness'. If materials are to be effective for learning they must be interactive and therefore must be structured so that such interaction can take place.

Structure and Layout

Hartley (1978) has pointed out that instructional materials are not intended for continuous reading and that

> the reader's focus of attention is constantly ranging from a place on the page to somewhere else: to another page, to the instructor, to the task in hand, to the chalkboard, to responses made by other learners, and, of course, back again to the place on the page.

It is argued, therefore, that printed instructional materials are used in an interactive and somewhat unpredictable sequence of events and that the spatial organisation of the text and its supporting material needs to provide a frame of reference within which the learner can move about without confusion.

Kozma *et al.* (1978) casually mention that layout and instructional aids may be valuable, stating that useful headings, subheadings, stated objectives and study questions, can help the student process and store information in the text.

What does the Research in the Field Indicate?

Hartley (1978) has suggested that headings and subheadings, together with a systematic use of space, convey more readily the structure of complex text, and Hartley and Burnhill (1976) that subheadings, or summary statements, placed in the left-hand margin, can help the reader to scan and to select relevant material. It has been claimed by Robinson (1961) that readers usually remember more when headings and subheadings are written in the form of questions rather than statements. Questions can, of course, be used other than as headings or subheadings, either at the start of a discourse or embedded in the text. Duchastel and Whitehead (1980) explain the rationale for in-text questions as a simple but powerful one: 'learning is maximised by an *active* information processing strategy which requires the learner to respond and at times reinterpret the information he or she is being provided with'.

The use of questions is believed to influence the depth of processing. Rickards and Di Vesta (1974) found that specific questions help people to remember specific cases and that higher-order questions lead to the recall of generalisations which included specific cases.

The statement of behavioural objectives at the beginning of instructional materials to help students organise their learning activities is fairly widespread, as evidenced by the Open University Science Faculty courses, and is probably particularly useful where the achievement test is based on well-specified objectives.

In a review of research relating to the usefulness of pre-instructional strategies in general, Hartley and Davies (1976) conclude that behavioural objectives are useful pre-instructional strategies, despite a general lack of agreement as to the level of detail to which they should be written, as the majority of studies indicate significant effects.

The idea of advance organisers as described by Ausubel (1963) is very much more difficult to use and to evaluate. Organisers are introduced in advance of the new learning material and are presented at what Ausubel describes as a 'higher level of abstraction, generality, and inclusiveness'. It is believed that the advantage of constructing a special organiser for each new unit of material is that the learner has the advantage of a subsumer which gives an overview of detailed material in *advance*, and provides organising elements that take into account the *particular context* contained in the material (Anderson and Ausubel, 1965). The major problem, according to Hartley and Davies is that there is currently no acceptable way of generating or recognising advance organisers, and the research on the effectiveness of advance organisers is certainly inconclusive.

The whole idea that people read straight through material from top to bottom has been questioned by Waller (1977), who has suggested that the reader is active, not only because he is struggling to understand what he reads, but also because he is selective in what he reads. He therefore regards the headings, organisers, etc., not as devices to assist comprehension, but rather as devices to help the reader find his way about and provide an access structure. Access devices are thus of two kinds: those which help the reader to *plan* his reading strategy (e.g. objectives, study guides) and those which help him *execute* his strategy (headings, section numbers, typographic devices). An extremely useful review of the structure of instructional text is that by MacDonald-Ross (1976).

In recent years there has been a considerable amount of research in the area of lecture handouts and student note-taking and the relationship between the two, which has been summarised by Hartley and Davies (1978) and reviewed by Howe and Godfrey (1978). It is clear that different types of handout result in different kinds of note-taking and, that between them, learning is likely to be affected.

Legibility and Layout of Print

Legibility of print can be regarded as the effects of typographical factors on the ease and efficiency of perception in reading. Tinker (1963) defines it thus:

> Optimal legibility of print is achieved by a typographical arrangement in which shape of letters and other symbols, characteristic word forms, and all other typographical factors such as type size, line width, leading (interlinear spacing), etc., are co-ordinated to produce comfortable vision and easy and rapid reading with comprehension. In other words, legibility deals with the co-ordination of those typographical factors inherent in letters and other symbols, words, and connected textual material which affect ease and speed of reading.

Type Style

There have been many studies of the legibility of different typefaces, and an investigation by Paterson and Tinker (1932) indicated that there is no significant difference in reading speeds between the common typefaces — except that normal 'typewriter face' gives a significantly lower reading speed.

A later investigation by Burt (1959), using a different experimental technique which recognised variation in both reading time and comprehension, indicated that reliable differences in the rate of comprehension could be found between what were previously regarded as 'good' typefaces. In this particular investigation the printing of scientific journals was being examined and Poulton (1972) has suggested that the superiority of the modern typeface in question could be accounted for by familiarity: 'scientists read most easily the kind of typeface to which they are used.' (p. 243) *Think about it — is your own preference for particular textbooks and journals related to the typeface used there?*

Variations in a Typeface

Typefaces can be printed in different weights (boldness), in different sizes and in italics. The research evidence indicates that italic type is read more slowly than ordinary lower case and that there is no difference in the speed of reading bold face and ordinary lower-case type. Nevertheless, it is usually suggested that italics and bold face should be used only for emphasis or for headings, as shown below.

Initiation of Translation

This encompasses the formation of the initial complex of ribosome, mRNA and amino acyl tRNA and the synthesis of the *first* peptide bond. Obviously, the presence of more than one amino acyl tRNA in the complex is required for the synthesis of a peptide bond. As noted earlier, the ribosome has *two* binding sites, the A and P sites. The A site has a high affinity for an amino acyl tRNA molecule but the P site has little or no affinity.

Letters can be printed in lower case (so-called 'small' letters) or upper case (capital letters). There is no doubt that lower case can be read more quickly than upper case (Tinker and Paterson, 1928). The explanation for this is that total word form is more important in perceiving words in lower case than in upper case, where perceiving occurs letter by letter.

It is well-established that reading by word units is a characteristic procedure of mature readers, and it is clear that this is facilitated by the distinctive word form to be found with lower case type.

Huey (1908, 1968) first pointed out that a fluent reader processes only part of what he sees, and that the upper part of a line of print is more informative than the lower part.

 porphyrin biosynthesis

porphyrin biosynthesis

More recent research (Kolers, 1968, 1969) has demonstrated that the right-hand side of the letters is more informative. *How often do you find yourself using capital letters in print, on the chalkboard, overhead projector, or slides?*

Type Size

The size of type alone does not determine legibility, and other typographical factors such as line width and interlinear spacing must be co-ordinated with type size. It appears, however, that with optimal line width and interlinear spacing, 9, 10, 11, or 12 point type may be used although it was found that reader preference was for 11 point. The main text of this chapter is set in 10 point.

11 point with 1pt leading

This encompasses the formation of the initial complex of ribosome, mRNA and amino acyl tRNA and the synthesis of the *first* peptide bond. Obviously, the presence of more than one amino acyl tRNA in the complex is required for the synthesis of a peptide bond. As noted earlier, the ribosome has *two* binding sites, the A and P sites. The A site has a high affinity for an amino acyl tRNA molecule but the P site has little or no affinity.

10 point set solid

This encompasses the formation of the initial complex of ribosome, mRNA and amino acyl tRNA and the synthesis of the *first* peptide bond. Obviously, the presence of more than one amino acyl tRNA in the complex is required for the synthesis of a peptide bond. As noted earlier, the ribosome has *two* binding sites, the A and P sites. The A site has a high affinity for an amino acyl tRNA molecule but the P site has little or no affinity.

10 point with 2pt leading

This encompasses the formation of the initial complex of ribosome, mRNA and amino acyl tRNA and the synthesis of the *first* peptide bond. Obviously, the presence of more than one amino acyl tRNA in the complex is required for the synthesis of a peptide bond. As noted earlier, the ribosome has *two* binding sites, the A and P sites. The A site has a high affinity for an amino acyl tRNA molecule but the P site has little or no affinity.

8 point with 1pt leading

This encompasses the formation of the initial complex of ribosome, mRNA and amino acyl tRNA and the synthesis of the *first* peptide bond. Obviously, the presence of more than one amino acyl tRNA in the complex is required for the synthesis of a peptide bond. As noted earlier, the ribosome has *two* binding sites, the A and P sites. The A site has a high affinity for an amino acyl tRNA molecule but the P site has little or no affinity.

12 pitch typewriter

This encompasses the formation of the initial complex of ribosome, mRNA and amino acyl tRNA molecules and the synthesis of the first peptide bond. Obviously, the presence of more than one amino acyl tRNA in the complex is required for the synthesis of a peptide bond. As noted earlier, the ribosome has two binding sites, the A and P sites. The A site has a high affinity for an amino acyl tRNA molecule but the P site has little or no affinity.

10 pitch typewriter

This encompasses the formation of the initial complex of ribosome, mRNA and amino acyl tRNA molecules and the synthesis of the first peptide bond. Obviously, the presence of more than one amino acyl tRNA in the complex is required for the synthesis of a peptide bond. As noted earlier, the ribosome has two binding sites, the A and P sites. The A site has a high affinity for an amino acyl tRNA molecule but the P site has little or no affinity.

10 pitch typewriter reduced to 70.7%

This encompasses the formation of the initial complex of ribosome, mRNA and amino acyl tRNA molecules and the synthesis of the first peptide bond. Obviously, the presence of more than one amino acyl tRNA in the complex is required for the synthesis of a peptide bond. As noted earlier, the ribosome has two binding sites, the A and P sites. The A site has a high affinity for an amino acyl tRNA molecule but the P site has little or no affinity.

12 pitch typewriter reduced to 70.7%

This encompasses the formation of the initial complex of ribosome, mRNA and amino acyl tRNA molecules and the synthesis of the first peptide bond. Obviously, the presence of more than one amino acyl tRNA in the complex is required for the synthesis of a peptide bond. As noted earlier, the ribosome has two binding sites, the A and P sites. The A site has a high affinity for an amino acyl tRNA molecule but the P site has little or no affinity.

Width of line

The acceptable width of line varies with size of type and interlinear spacing. With 10 point type, set solid (i.e. no additional spacing

Width of Line

The acceptable width of line varies with size of type and interlinear spacing. With 10-point type, set solid (i.e. no additional spacing between the lines) it appears that line widths between 75 mm and 115 mm (line widths are usually measured in 'picas', a pica being about 4.2 mm) or 17–27 picas are acceptable. Increasing the interlinear spacing increases the range of line length which is readable. The text of this chapter is set in 10pt Press Roman with a line width of 24 picas and additional interlinear spacing of 2 points. *In typed instructional materials which you use with your students is the line width within the acceptable limits?*

Type size, interlinear spacing and line width are interrelated factors affecting legibility of typographical arrangements. Apart from their legibility, very short and very long lines, and material set solid are disliked by readers.

Justified and Unjustified Typesetting

Most books and journals are printed with 'justified' typesetting which means that a straight right-hand margin has been achieved by varying the spacing between words and by sometimes using hyphenation. If regular spacing between words is used the lines of print will be of varying lengths (as in typescript), and this is known as 'unjustified' typesetting. Unjustified typesetting usually contains fewer hyphenations.

There has been considerable controversy over justification concerning legibility, instructional value, cost and aesthetic appeal. Experiments investigating the instructional value of justified and unjustified settings have been carried out by Zachrisson (1965), Spencer (1969) and Hartley and Mills (1973). For adult readers, Hartley and Mills found:

(1) no difference in speed of reading between justified and unjustified text;
(2) no difference in comprehension;
(3) no difference in preference.

The increasing use of word processors for setting instructional text in higher education institutions may well lead to an increased use of justified copy as these machines can readily produce material in justified form.

Size and Shape of the Page

Hartley (1978) chooses to use a discussion of page size for the opening chapter of his helpful book on designing instructional text.

> Decisions about page size are crucial because they form the baseline from which the remaining typographic decisions are made. Once the size of the page is determined, it is next possible to decide upon the layout of the page, the interlinear spacing, the line-lengths of the text, the position of illustrations and so on. In other words, page-size is to typographic planning as site-size is to the design of buildings: it places manageable limits on what can be done by way of arranging sensibly the prefabricated parts.

A major difficulty in planning this particular chapter has been the size of the page which, although allowing an acceptable line width, does not allow anything other than a 'side-to-side' arrangement of the text. In other words, it is impossible to plan the page on a grid system and take decisions regarding alternative layouts as the line width would become absurdly short. However, most instructional materials which are produced 'in house' are likely to be produced on larger formats, in particular on A4 (210 mm x 297 mm) the best known of the ISO A series of sizes.

A major advantage of the ISO A series is that the sides are in the ratio of $1:\sqrt{2}$ and if halved or doubled the ratio remains the same. Thus it is possible to produce from one A3 sheet (i.e. two sheets of A4 side by side) an A4 sheet (in effect, two sheets of A5 side by side) by reducing to 70.7 per cent of original size. This reduction feature is characteristic of many photocopiers and automatic duplicating equipment although many of them, while claiming to reduce A3 to A4 operate a reduction to 66 per cent (this being appropriate to American sizes) and in the process destroy the $1:\sqrt{2}$ ratio.

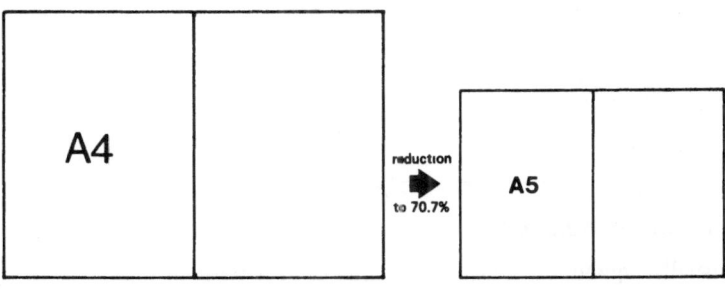

An important consideration regarding the rationalisation of page sizes is in the relationship between how the material is intended to be used and the way in which its component parts can be arranged on the page. Hartley (1978) has argued that the recognition of standards by research workers is a necessary condition for further development in the design and evaluation of instructional text.

For a detailed discussion of planning the layout of instructional materials the reader is referred to Chapter 2 of Hartley's *Designing Instructional Text,* but the primary guidelines for layout can be derived from the typographical and educational principles outlined earlier in this chapter.

Assuming that an A4 vertical configuration is going to be used and the typeface is about 10 or 11 point, it is clear that the maximum length of line which would be within the acceptable legibility standard discussed earlier is about two-thirds of the width of the page. The possibilities are, therefore, to have a double-column page (as in many journals) or an asymmetrical arrangement with the body of text utilising only two-thirds of the page. The discussion regarding headings, subheadings, advance organisers and access structure suggests that there is value in having readily identifiable key phrases, particularly in the left-hand margin.

From both typographical and educational points of view, therefore, it would appear that the best arrangement of material on the page is to divide the page vertically in three, putting headings and subheadings in the left third and the body of text in the right two-thirds.

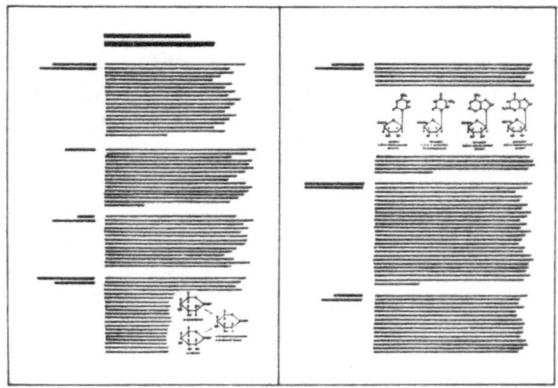

What about Illustrations?

It is usually argued that illustrations are used for two purposes:

information and motivation (Magne and Parknas, 1963); but Vernon (1953) points out that the motivational effect of pictorial material varies greatly with age, intelligence and education. Smith (1960) argues that the use of impressionistic, representative and abstract illustration brings effective organisation to textbooks, and creates books which not only meet the perceptual and artistic needs of the student, but also refine his understanding of what is read, improve his retention of material and stimulate the development of creative thought. He states that creative art in books serves three main functions.

> First, it serves to *perceptually motivate* the reader – to attract him to pick up the book, to explore it, and above all to develop a high feeling of expectancy in turning each page. Second, artistic illustration *perceptually reinforces* what is read, so that the situations, events and relationships described in words are made more meaningful and thus are better retained. Finally, correlated art *symbolically enhances* and deepens the meaning of the verbal material and thus serves to advance organisation of the verbal materials to promote creative thinking. (p. 29)

A somewhat similar approach has been proposed by Duchastel (1978) who describes the three roles of illustrations in text as attentional, explicative and retentional, claiming that the first two are fairly intuitive and commonly found in instructional texts, and acknowledging that the third one is less intuitive and more speculative. The basis of the retentional role is that iconic or pictorial memory is less resistant to forgetting than is verbal memory, and that illustrations used in this role are presumed to act similarly to section headings, forming a conceptual plan of the subject matter for the learner.

But perhaps the most useful research into the use of illustrations (in print and also in other media) has been that carried out by Dwyer (1972) and summarised in his helpful book *Improving Visualized Instruction.* Dwyer's main concern has been the degree of realism needed in illustrations and it is interesting to note how effective simple line drawings can be in facilitating the achievement of certain kinds of learning objectives.

Concluding Remarks

The production of instructional text can be greatly improved through consideration of research findings relating to both student learning and typographical design. A recent publication by Hartley and Burnhill (1978), *Fifty Guidelines for Improving Instructional Text*, is a very helpful summary.

In the production of printed instructional materials for your students, to what extent have you considered the research findings?

9 COMPUTER-BASED LEARNING IN BIOCHEMISTRY

Charles F. A. Bryce

Introduction

The origins of the application of digital computers to the process of
instruction can be traced back essentially to the pioneering work on the
'teaching machine' and the development of programmed learning by
such workers as Thorndike, Skinner, Pressey and Crowder (Coulson,
1962; Glaser, 1965; Bushnell and Allen, 1967; Stolurow, 1967;
Atkinson and Wilson, 1968; Skinner, 1968) although it has been
argued that this really only represents the hardware side of things and
that the fundamental rationale has its roots as far back as Plato and
Aristotle (Blaisdell, 1976). With reference to science education in
particular, the application is also attributed in part to developments in
the use of computers in the subject of science itself in, for example, the
processing of experimental data, computer simulation studies and
problem-solving (Hooper, 1978).

It is fair to say that such an application of computers and the
associated hardware to the process of instruction in the period from
about 1960 onwards has been, and indeed still is, a somewhat
controversial and emotive issue. However, without wishing in any way
to add to the already vast literature dealing with the pros and cons for
its implementation, it was felt that a brief account may serve as a
useful basis from which to develop, and to comment on, a number of
topics to be considered later in this chapter. For this purpose, Figure
9.1 has been constructed and summarises the reported views of the
advantages and disadvantages of the use of computers in higher
education as expressed by a number of workers in the educational
technology field, although the list is in no way exhaustive.

Another index of non-conformity in the use of computers in
education can be seen in the terms used to describe the application, for
example, computer-managed instruction (CMI), computer-managed
learning (CML), computer-assisted instruction (CAI), computer-assisted
learning (CAL), computer-based learning (CBL), plus a number of
other equally appropriate terms and acronyms. It would appear from
common usage that *computer-based learning* is the all-encompassing

Figure 9.1: Reported Views on the Advantages and Disadvantages in the Use of Computers in Higher Education

Advantages in the Use of Computers in Higher Education
Permits students to study multivariate systems with immediate individual feedback (Hooper, 1977)
Reduces the time required by students to carry out simple or complex calculations, to access long data bases and to analyse data, so allowing more time on other aspects of teaching (Crovello, 1974; Hooper, 1977)
Provides students with the opportunity of experiencing experiments by simulating what might otherwise be dangerous, costly, seasonal, unrealiable or time-consuming events (Ayscough, 1973; Crovello, 1974; Hooper, 1977; Smythe and Lovatt, 1979)
Generates familiarity with computer use and this is seen as an advantage in later pursuits (Hooper, 1977)
Can relieve the lecturer of some administrative burdens (Hooper and Toye, 1975)
Good for aggressive students (Blitz, 1972)
Permits student-based individualised instruction (Suppes, 1967; Atkinson and Wilson, 1968; Crovello, 1974; Cross, 1976; Hooper, 1977)
Facilitates course syllabus development (Weinberg, 1979)
Offers impersonal teaching situations for students to make mistakes without any public embarrassment (Hooper, 1977)
Increases student motivation (Crovello, 1974; Hooper, 1977)
Provides students with essentially drill and practice opportunities to consolidate/reinforce material (Suppes, 1967; Hooper and Toye, 1975; Hooper, 1977; Leiblum, 1977)
Can generate, mark and analyse tests (Marmion and Lutz, 1971; Wood, 1975; Hooper and Toye, 1975; Cross, 1976; Cnop-Grandsard, 1979)
Store and update class records (Broderick and Lovatt, 1975; Hooper and Toye, 1975)
Humanizes (personalizes) (Miller, 1979)

Figure 9.1 – *continued*

Disadvantages in the Use of Computers in Higher Education
Instructors may feel threatened (Grimm, 1978)
Students and staff must be given special training to work in the system (Glaser, 1968; Crovello, 1974)
Excessive time demands on course authors to write programs and appropriate documentation (Mitzel, 1966; Leiblum, 1977; Moore and Collins, 1979)
No theoretical basis (Gentile, 1967)
Lack of compatability between systems (software and hardware) and hence difficulty in transferring programs (Mitzel, 1966; Moore and Collins, 1979)
Often tedious to anticipate particular student responses when programming (Wallace, 1978)
Little control of the reliability of the computer system and access (Wallace, 1978; Grimm, 1978; Smythe and Lovatt, 1979)
Can hinder the development of analytical skills if overused (Crovello, 1974)
Since it is for individualising the learning, really need one terminal per student (Broderick and Turner, 1975)
Poor for orderly, deferent and endurant students (Blitz, 1972)
Less efficient for high anxiety students (Rappaport, 1971)
Expensive (Gerrick, 1971; Smythe and Lovatt, 1979; Rushby, 1980)
Response time can be slow, especially with time-sharing systems (Wallace, 1978)
Dehumanizing (Wallace, 1978)

term whilst that described as *computer-assisted* refers to the use of the computer as a means of presenting material to the learner, posing questions, analysing and evaluating the answers, directing from one part of the program to another in response to the learners' current progress etc., whilst the term *computer-managed* is reserved to describe the use of the computer as a means of generating, marking and/or analysing tests, storing class records and fulfilling other administrative functions (Moore and Collins, 1979; Smythe and Lovatt, 1979; Harding, 1980).

There have been descriptions given to differentiate the terms computer-assisted instruction and computer-assisted learning which rely largely on a consideration of whether or not a tutorial mode is present (Hooper and Toye, 1975; McKenzie *et al.*, 1978; Smythe and Lovatt, 1979). Having said that, it would appear that often this distinction is not recognised and that the two terms are used interchangeably. This may or may not represent an individual's own dogma of the education process as being either a product view in which learning is the outcome, or a process view since instruction has been defined as 'a set of events designed to initiate, activate and support learning' (Gagné and Briggs, 1974). For the purposes of the present chapter this distinction will be overlooked and the terms computer-managed learning (CML) and computer-assisted learning (CAL) will be used throughout. To avoid unnecessary verbosity, the word computer used hereafter will be used collectively to include main-frame computers, minicomputers, microcomputers and microprocessors.

It is neither the intention nor the purpose of the present chapter to provide detailed accounts of the fundamentals of either computer programming or computer architecture and machine configuration, but rather to review the developments that have emerged in the last few years particularly where these are specifically related to the subject of biochemistry. For those readers not too familiar with the computer hardware aspect, two very readable and comprehensive accounts are given by Moore and Collins (1979) and Moore *et al.* (1979), whilst those requiring a knowledge of computer programming skills are referred to texts by Monro (1978) and Huntington (1979) for BASIC, and Orr *et al.* (1973) or Beech (1975) for FORTRAN, together with the appropriate user manual for their own computer and thereafter plenty of 'wet-hands' experience. For those readers wishing a more detailed and general background to computer-assisted learning, a particularly useful source is a recent article by Harding (1980) in which he describes the recent developments in CAL by considering nine independent books or reports in the field. Also of value in this respect is the recent book by Rushby (1979). As far as a choice of computer language goes, then this is summarised in Figure 9.2.

Figure 9.2: Commonly-used Programming Languages

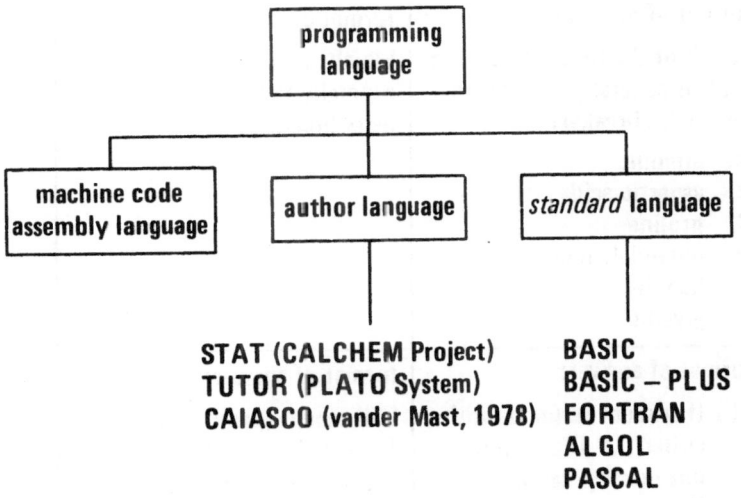

<pre>
                    STAT (CALCHEM Project)   BASIC
                    TUTOR (PLATO System)   BASIC – PLUS
                    CAIASCO (vander Mast, 1978) FORTRAN
                                              ALGOL
                                              PASCAL
</pre>

Authors of CAL-packages would not normally be involved in writing a program in machine code but may, in the case of a microprocessor, wish to make use of some assembly language. More often than not, particularly with transient users, the choice of language will be one of the higher-level languages which can be either an author language or a *standard* language. The principal advantage of the author language is that it can easily cope with completely free format responses from the user whereas this would be difficult with, for example, FORTRAN or BASIC. As an example, consider the two approaches outlined in Figure 9.3.

Clearly, the advantage of the standard-language format is that it is slightly easier to program than the author-language version but the disadvantage is that the student is getting some assistance with the recall of knowledge. The disadvantage of author languages is the problem of catering for spelling variations, word order, double negatives and word mismatching. FORTRAN and BASIC have the advantage of being compatible with many computers (with slight variations) and so programs written in these languages can be transferred/transported relatively easily.

Figure 9.3: Protocols for Formatting Questions and Answers in Computer-assisted Learning Using Different Programming Languages

BASIC/FORTRAN	author language
format of question: **Which of the following amino acids is generally thought to be a helix breaker?** **A. arginine** **B. aspartic acid** **C. proline** **D. phenylalanine** **E. histidine** **F. glycine**	**format of question:** **Which amino acids are generally thought to be helix breakers?**
format of answer: (i) **If student responds with either A,B,D or E put out message saying "Wrong answer, try again."** (ii) **If student responds with C put out message saying "Yes, absolutely correct, although proline helices can and do exist in proteins".** (iii) **If student responds with F put out message saying "Yes, but really only if present in the protein in high concentration".**	**format of answer:** **Will accept (pro)line and also (gly)cine), (ala)nine), (ser)ine), (thr)eonine) or (val)ine) in (high) (conc)entration)** **[in any order]**

Computer-assisted Learning

Nearest-neighbour Frequency Analysis (NNFA)

A number of detailed aspects in the design, production and validation phases of a CAL-package on the technique of NNFA have already been considered in Chapter 1. It is the intention of the present section to discuss this package in a little more detail particularly from the point of view of the programmer and the user.

NNFA was a technique which was initially devised to provide conclusive proof that DNA polymerase I was causing template-directed DNA synthesis rather than a random polymer of the constituent deoxyribonucleoside 5'-triphosphates. There was at this time a considerable amount of supportive evidence that this was indeed the case, but NNFA proved to be the first piece of experimental evidence that was conclusive (Kornberg, 1961; Josse *et al.*, 1961). Almost as a bonus, the technique provided in addition the first experimental proof that the DNA duplex strands have opposite polarity and the Chargaff's base equivalence can be demonstrated. A summary of the experimental protocol is given in Figure 9.4.

Plainly then, with such a powerful technique it is important that a very serious effort is made to help the students to comprehend fully the technique and its potential application. Already appreciating those areas of this topic which have, in the past, proved difficult or confusing to the student, due stress and attention were given to these in the nearest-neighbour frequency analysis simulation (NNFSIM) CAL-package. Figure 9.5 illustrates a typical run with the NNFSIM program and should give the reader a feel for the package as a whole although fairly large parts of the responses etc. have been edited out to make it more precise.

As we mentioned earlier, the package is composed of three distinct phases, an introduction, an assessment and a simulation section. The introduction can be a summary of the technique of NNFA and/or managing instructions for the package itself. For users' responses of yes or no, the program is written to accept YES, Y, NO or N in either upper- or lower-case letters as can be seen in Figure 9.5.

The assessment section contains ten multiple-choice questions ranging from true/false to some with six distractors included. In response to the user's input, immediate feedback is given and this is either of encouragement for a correct response or of prompting for an incorrect answer. In this package students are cycled around questions answered incorrectly until such times as they obtain the correct answer. For other packages (see later) the number of cycles was restricted to two before going on to the next question. The assessment section in this program is linearly structured and so all users are processed through the same ten questions following which the score, expressed as a percentage, is output on the user's terminal. The third phase, and

Figure 9.4: Experimental Protocol for Nearest-neighbour Frequency Analysis

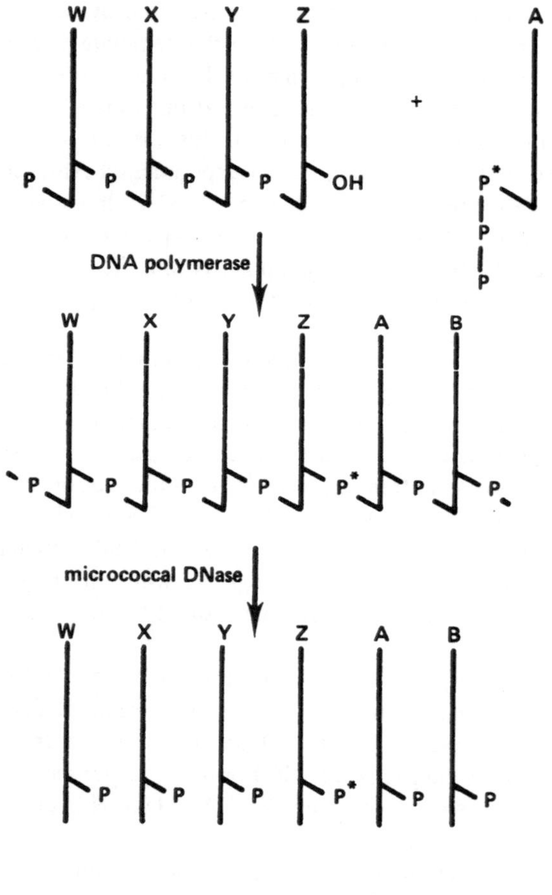

— measure frequency of label in each deoxyribo-nucleoside 3'-monophosphates

— repeat procedure three more times using a different labelled triphosphate each time

Figure 9.5: Sample Run with NNFSIM Program

Welcome to Nearest-Neighbour Frequency Analysis
****

Do you wish to see an Introduction?
(Type in yes or no, then depress return)

?Yes

This program is described as interactive in that, at various points within the program, questions are raised and answers are expected. The form of the expected reply is fairly obvious, for example:

Do you wish . . . .

You would reply by typing in 'yes' or 'no' (or 'y' or 'n').

Similarly, for an objective test item with say 4 choices of answer:

A ....
B ....
C ....
D ....

You would answer by typing in the letter of your choice.

Note, however:
****

Everytime you respond interactively, depress the key marked 'Return'
******

Good luck with the program and I hope you learn something worthwhile!

————————————————————————————————————

Do you wish to be assessed in the procedure   ?y

Objective Tests in Nearest-Neighbour Frequency Analysis (NNFA)
****

Question 1

During DNA-dependent DNA biosynthesis, the nucleotide sequence in the template DNA is reproduced in the product.

A   True
B   False

?A

Well done . . . . now try question 2

Figure 9.5 – *continued*

Question 2

Which of the following is not essential for DNA-dependent DNA biosynthesis.

       A   All four deoxyribonucleoside 5'-triphosphates
       B   magnesium ions
       C   DNA template
       D   DNA-dependent DNA polymerase
       E   Sigma factor

        ●
        ●
        ●
        ●
        ≘

Well done . . . . that is the end of the objective assessment.

      Your score for the objective test section was   100%
      ****

———

Now let us see the results of a computer-simulated experiment
----

          Nearest-Neighbour Frequency Analysis
          ****

       Exercise using computer-simulated experimental data

        Sequence of native DNA primer is as follows:

  5'   3'
    CCTTTGGCTTAATACCAAGGCTGGACCTGATTCAG
    GGAAACCGAATTATGGTTCCGACCTGGACTAAGTC
3' or 5'   5' or 3'
                                       ****
                                       Return
                                       ****

Figure 9.5 – *continued*

Dinucleotides		Exptl. Data	
	*		*
A-A * G-A		788	* 386
A-G * G-G	*	692	* 773
A-C * G-C		389	* 782
A-T * G-T	*	595	* 489
***********	*	***********	*
T-A * C-A	*	382	* 784
T-G * C-G		800	* 198
T-C * C-C		387	* 792
T-T * C-T	*	860	* 666

```
*******
*Return*
*******
```

?

Are you using a VDU (type VDU) or a teletype (type TTY) . . . . . . ?VDU

- •
- •
- •
- •
- •

Studying the sequences, would you say that the DNA was synthesised randomly or template-directed?

If random, type R; if template-directed, type T

- •
- •
- •
- •
- •

```
***************************************************************
```

If you wish to go through the program (or part of the program) again, then type again, otherwise type end . . .

perhaps the most important, generates a truly random set of simulated experimental data on which the user makes a number of deductions. For some of these, help in the form of prompts is available to assist the student in making a decision and giving a particular response. As Hartley and Lovell (1977) pointed out, to make scientific phenomena accessible to the intellect requires the teacher to illustrate and build up from the underlying principles. The use of simulation as a means of providing a 'working model' has been shown to be extremely effective in this respect. Applying Skinner's educational psychology principles, the provision of instant feedback should act as a powerful reinforcing stimulus; however, the value of such reinforcement has never been demonstrated (Grundin, 1969). Hartley and Lovell provide very useful recommendations regarding the optimal pace of CAL-packages and the nature of the feedback to be provided.

Enzyme Kinetics

Enzyme kinetics has always been a topic that is appropriate for study by computational methods partly because of the interest in processing kinetic data by non-linear methods or by a number of different linear transformations and also in simulating a variety of enzyme mechanisms (Wilkinson, 1961; Cleland, 1963; Wieker *et al.*, 1970; Atkins and Nimmo, 1973; Kuhn and Brand, 1973; Cornish-Bowden and Eisenthal, 1974; Roberts, 1977; Fukagawa *et al.*, 1980). This interest is reflected in a number of CAL-packages for undergraduate student use being published in the last few years (Friedland and Rosen, 1971; Orr *et al.*, 1973; Heydeman, 1977; Hancock, 1978; Essenberg, 1980; Pinnick, 1980).

Of these, the two which I have had most experience in operating with undergraduate classes are ENZKIN (Heydeman, 1977) and LOCKEY (Friedland and Rosen, 1971) and these have proved eminently successful in providing students with model multivariate systems which they can manipulate with ease and derive their own conclusions from the results. For this topic, such experience plays a particularly useful role since it forms a meaningful bridge between the associated lecture program which tends to concentrate more on theoretical aspects of the topic and the associated practical program where, at least initially, the emphasis is on technique and accuracy. The output format from both of these learning packages can be preselected to either be in the form of a table or as a line-printer plot. The first option makes use of significantly less computing time and also provides students with first-hand experience of plotting their own data and drawing conclusions from them. Alternatively, if neither of these activities is of prime importance, then students can select a line-printer plot which, although slower in response, does allow them to draw conclusions quickly and make any alterations to the system as they think fit on the basis of the information available to them. This relatively slow response time is much less obvious to students using a visual display unit (VDU) where the speed of information display is increased by a factor of about 20–100 (i.e. baud rates of 2,400–9,600 compared with 110 for the conventional teletype). The disadvantage with the VDU for this type of work is that the image is transient, in that once a new set of data is input we lose the previous plot from the screen. If this proves to be a problem then there are a number of solutions. For example, if a main-frame computer is being used then it may have a system command (e.g. @ PHOTO) which creates a copy of what is taking place at a particular terminal and this can be output on the line printer. Alternatively, a photograph of the plot etc. on the VDU screen can be taken very easily using a Polaroid hand-held camera with attached hood (CU-5). Examples of the two types of output format are given in Figure 9.6 and from an interpretative point of view the line-printer plot is very much more easily reconcilable – a picture is worth a thousand words, as pointed out by McKenzie *et al.* (1978).

In the case of ENZKIN and LOCKEY, the quality and resolution of the line-printer plots are adequate for the educational intentions of the packages. For other applications, however, this is certainly not the case and this necessitates recourse to high-quality, high-resolution graphics terminals as will be described.

Recently, whilst validating a student workbook on graphical procedures in enzymology which has been used in this college by a considerable number of students over a two-to three-year period, it appeared that in achieving one of the objectives the students were almost blindly accepting what was written in the booklet without questioning it or thinking about it in anything but the most tentative fashion. This particular objective related to the merits and limitations of a hyperbolic plot and to the associated linear transformations of the Michaelis-Menten equation. In order to provide the student with more actual experience of handling kinetic data, and hopefully in the process augmenting their skills of critical analysis, each was supplied with a set of experimental [S] and v values which they input into a graphics package designed specifically for this study (Bryce, 1981a). This

Figure 9.6: Sample Output from the LOCKEY Program Illustrating the Alternative Output Formats

```
************************************************************

AMOUNT OF ACETYLCHOLINE - FROM 0 TO 3 MILLIMOLES ?2
TYPE OF INHIBITOR - USE CODE FROM 0 TO 5 ?2
AMOUNT OF INHIBITOR IN MILLIMOLES ?120
DATA FORMAT: 1=TABLE, 2=GRAPH ?1

MINUTES       ACETYLCHOLINE    TOTAL ACETIC
ELAPSED       REMAINING        ACID PRODUCED
-------       -------------    ------------
0             2                0
0.1           1.77             0.23
0.2           1.55             0.45
0.3           1.34             0.66
0.4           1.14             0.86
0.5           0.95             1.05
0.6           0.77             1.23
0.7           0.61             1.39
0.8           0.47             1.53
0.9           0.35             1.65
1             0.25             1.75
1.1           0.17             1.83
1.2           0.12             1.88
1.3           0.08             1.92
1.4           0.05             1.95
1.5           0.03             1.97
1.6           0.02             1.98
1.7           0.01             1.99
1.8           0.01             1.99
1.9           0              2
THE REACTION HAS RUN TO COMPLETION

CONCENTRATION OF INHIBITOR REMAINING: 120 MILLIMOLES

ANOTHER EXPERIMENT? (1=YES, 0=NO) ?1

************************************************************

AMOUNT OF ACETYLCHOLINE - FROM 0 TO 3 MILLIMOLES ?2
TYPE OF INHIBITOR - USE CODE FROM 0 TO 5 ?2
AMOUNT OF INHIBITOR IN MILLIMOLES ?120
DATA FORMAT: 1=TABLE, 2=GRAPH ?2
```

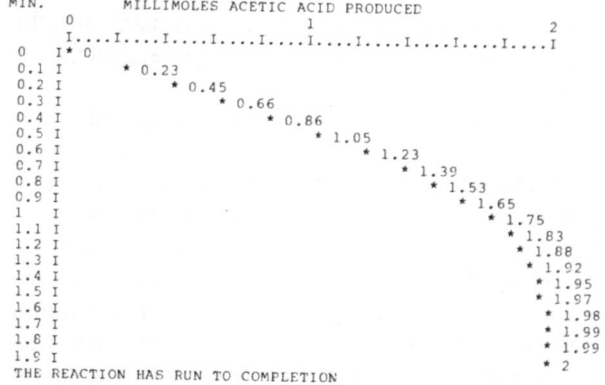

```
MIN.         MILLIMOLES ACETIC ACID PRODUCED
      0                        1                          2
      I....I....I....I....I....I....I....I....I....I....I
0     I* 0
0.1 I      * 0.23
0.2 I         * 0.45
0.3 I            * 0.66
0.4 I               * 0.86
0.5 I                  * 1.05
0.6 I                     * 1.23
0.7 I                        * 1.39
0.8 I                         * 1.53
0.9 I                           * 1.65
1   I                             * 1.75
1.1 I                              * 1.83
1.2 I                               * 1.88
1.3 I                                * 1.92
1.4 I                                 * 1.95
1.5 I                                 * 1.97
1.6 I                                  * 1.98
1.7 I                                   * 1.99
1.8 I                                   * 1.99
1.9 I                                    * 2
THE REACTION HAS RUN TO COMPLETION

CONCENTRATION OF INHIBITOR REMAINING: 120 MILLIMOLES

ANOTHER EXPERIMENT? (1=YES, 0=NO) ?1
```

program analysed the data and output high-quality, partially-labelled graphs containing the data points plus the best straight lines through them (in the case of the linear transforms). The plotting procedures adopted were the Lineweaver-Burk plot, the Eadie-Hofstee plot, the Woolf plot, the direct-linear plot and the Michaelis-Menten plot (Michaelis and Menten, 1913; Woolf, 1932; Lineweaver and Burk, 1934; Eadie, 1942; Hofstee, 1952; Cornish-Bowden and Eisenthal, 1974). A typical set of graphic output is shown in Figure 9.7.

From such data, the student is required to provide suitable labels for the various axes, evaluate the kinetic parameters V_{max} and K_m, consider the data points in relation to the best straight line (evaluated by a linear least-squares regression routine) and to each other in the case of the linear transforms. The Michaelis-Menten plot should illustrate some of the difficulties in working with non-linear plots, particularly with respect to determining the asymptote to the curve (tendency is always to put it too close to the curve). In this case, the program can be slightly modified to plot the hyperbola without the asymptote. In all cases the estimated values of V_{max} and K_m from computations written within the program itself are output separately on the on-line line printer. In this way the student can be supplied with the graph output and his evaluations etc. checked against the answers generated by the program.

In the very short time that this package has been available to students it has only been possible to undertake a cursory formative evaluation which has been rather subjective. The signs are, however, that this approach will form a very useful adjunct to the lecture and practical courses on enzyme kinetics and for this reason it was felt to be a profitable exercise. The one important drawback related to CAL-graphics packages is their poor transferability since many of the graphics software packages are machine/device-dependent. However, providing the program is not too lengthy and is well-documented, then it should not be too demanding a task to modify the program accordingly.

Another recent example of the way in which high resolution graphics can significantly improve a non-graphics CAL-package can be seen in a study dealing with a variety of rapid nucleic-acid sequencing methods (Bryce, 1981b) as illustrated in Figure 9.8.

There is also a graphics version of ENZKIN available as part of the CUSC (Computers in the Undergraduate Science Curriculum) Project (McKenzie *et al.*, 1978). One very large program that has been developed in the area of CAL is the PLATO system at the University

Figure 9.7: Sample Output from the High-resolution Graphics Program on Graphical Procedures in Enzymology

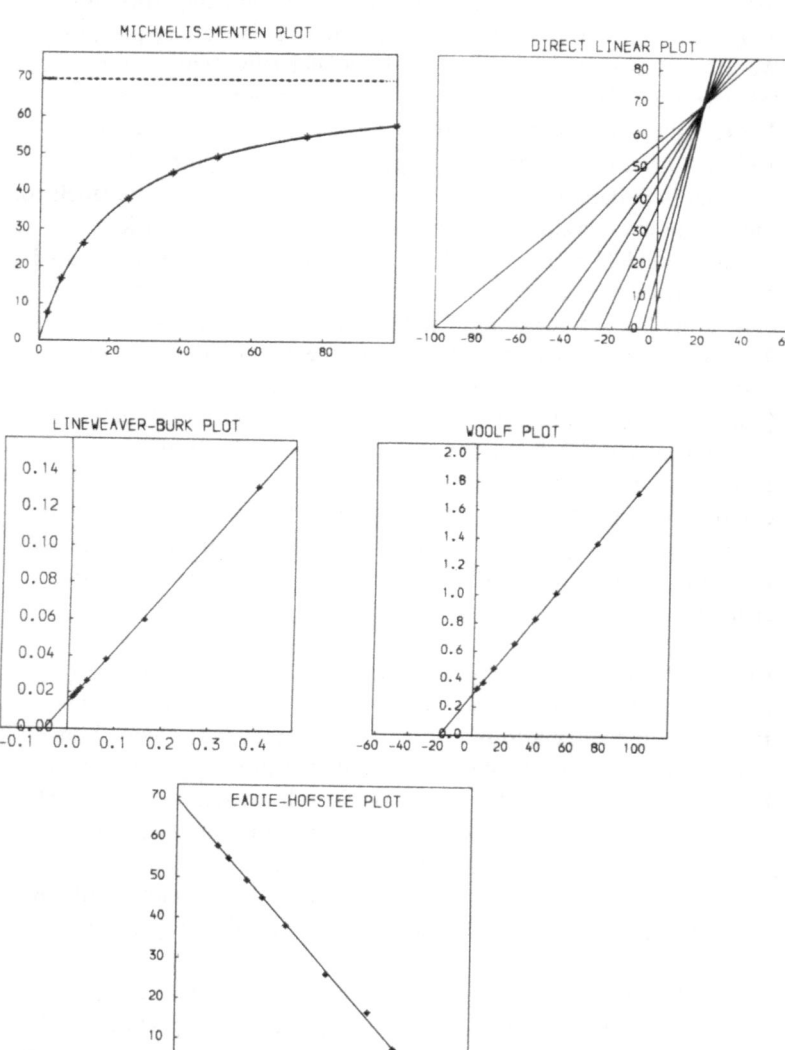

Figure 9.8: Sample Graphics Output Using (a) Low-resolution Graphic Facilities on a Conventional Teletype or VDU and (b) High-resolution Graphics Facilities on a Storage Screen Terminal

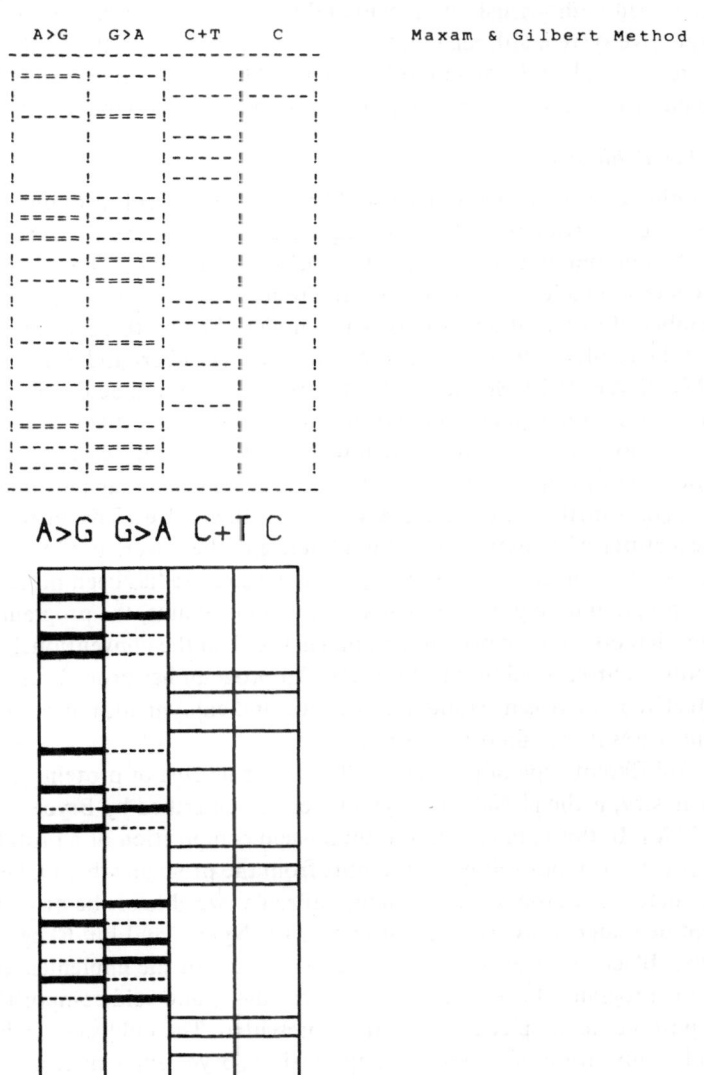

MAXAM AND GILBERT METHOD

of Illinois and this has a number of biochemistry lessons incorporated (Hyatt *et al.*, 1972; Smith and Sherwood, 1976). This also has a number of additional features such as touch-sensitive screens and interfaced audio-visual equipment and for these reasons is an extremely expensive system although links are being made outwith the United States. A rival to this mammoth system is TICCIT which uses modified colour TV sets to present computer-assisted and video-tape lessons.

Other Topic Areas

Another area which lends itself well to CAL treatment is protein sequence analysis or nucleic-acid sequence analysis. Here, use is made of the random-number function to generate a polymer sequence of a predetermined length and this information is stored in the memory. A number of different operations can be undertaken on this sequence and the results supplied to the user's terminal (Daubert and Sontum, 1977; Bryce, 1977; Shone, 1979). In this way it should be possible to characterise the sequence in part, if not in total. Daubert and Sontum (1977) noticed that their program was effective in both teaching the rationale of protein sequence analysis and also in encouraging students to request further information and so teach themselves. This provided the lecturer with 'extra' curriculum hours and these were used to explore the subject more fully. Again, extensive use has been made of this program by my own students and, without doubt, the program is considerably more expedient and pragmatic than the conventional lecture course. In addition, this particular program has proved very effective in increasing student motivation and enjoyment, indeed in some cases it was almost too successful.

A different type of program, still in the same area of protein chemistry, is the MINMWT program recently described by Bryce (1979c). In this application, the amino-acid composition of a protein or peptide is input following prompts from the program which takes this data and outputs the minimum molecular weights of the protein/ peptide under study using a best integer fit (Nyman and Lindskog, 1964; Black and Hogness, 1969). As an example of the implementation of the program, the experimental data for the amino-acid composition of purified horse-spleen apoferritin is presented. The published amino-acid composition of horse-spleen apoferritin (Bryce and Crichton, 1971a) was used and from Figure 9.9 it can be clearly seen that a minimum value of the fraction of maximum deviation for fit, f, occurs at a molecular weight value in the range 19,000–19,400. By using smaller increments it was possible to demonstrate that the actual

Figure 9.9: The Relationship between the Molecular Weight of Horse-spleen Apoferritin and the Fraction of Maximum Deviation, f, of the Amino-acid Frequencies

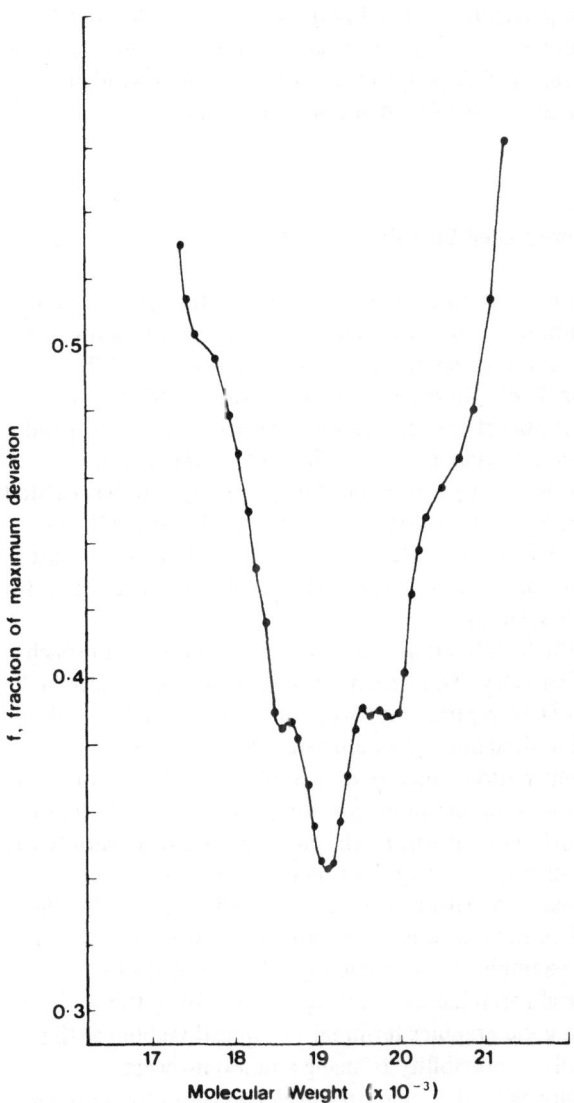

Source: Bryce, C. F. A. (1979) *Laboratory Practice*, **28**, 403

minimum occurred at a value of 19,280. The amino-acid composition, corrected to the best integer fit, is output on request as shown in Figure 9.10.

The minimum molecular weight obtained from the amino-acid composition in this particular case, 19,280, agrees very well indeed with estimates from a variety of physical and chemical methods (Bryce and Crichton, 1971a, b). This program has been used with student projects and also with a class of students studying analytical biochemistry.

Improved Computer-assisted Learning

Recently a number of workers have extended the potential of CAL by interfacing the computer system with addition hardware. Examples of this type of application include interfaced tape-film (Fox, 1979; Branson and Foster, 1980), interfaced audio (Hertzler, 1979), interfaced random-access microfiche (Bryce and Stewart, 1979, 1980) and interfaced random-access videodisc (Branson and Foster, 1980; Andriessen and Kroon, 1980). Although this type of system has existed with respect to the very large-scale systems like PLATO and TICCIT, these more recent applications differ in that they are designed for use by the large majority of lecturers using CAL who do not have access to such large computer systems.

An example of the benefits that can be derived from this approach in the field of biochemistry can be seen by considering an assessment/tutorial package on DNA replication (Bryce and Stewart, 1979, 1980). We were interested in designing a student self-assessment package for the topic of DNA replication which had, in addition to the various test items, a series of tutorial modes built into the program as a whole. In order to cater for different student levels and also for a wide variety of routes through the package (largely dependent on current performance), computer-assisted learning was chosen as the medium for the package. The problem then, however, was how to reproduce complex visual material, for example electron micrographs, detailed line drawings or structural formulae within the program. Since the resolution of even a good graphics terminal was questionable for this work, we considered the possibility of using a random-access projection system alongside the computer package. From the number of random-access projection systems available commercially, we chose the Revox Audiocard System which uses a 5 x 12 grid colour

Figure 9.10: The Refined Integer Fit of Amino-acid Residues Derived from the Experimentally Determined Amino-acid Composition of Horse-spleen Apoferritin

```
**************************************************

            AMINO ACID COMPOSITION

    LYSINE ................................................  9
    HISTIDINE ..........................................  6
    ARGININE ...........................................10
    ASPARTATE ........................................18
    THREONINE .......................................  6
    SERINE ...............................................  9
    GLUTAMATE .....................................25
    PROLINE .............................................  3
    TRYPTOPHAN ....................................  2
    ALANINE .............................................15
    CYSTEINE ...........................................  3
    VALINE ................................................  7
    METHIONINE .....................................  3
    ISOLEUCINE ........................................  4
    LEUCINE..............................................26
    TYROSINE ...........................................  5
    PHENYLALANINE ..............................  8
    GLYCINE .............................................10

**************************************************
```

microfiche, partly because of its high speed of frame change, its ergonomic design and its extra features like lamp off, lamp on, left-hand side of screen illuminated and right-hand side of screen illuminated. This microfiche system is operated by depressing the relevant keys in a small control box. The next stage of this study was to design an interface which, by using computer software, would allow us to bypass the control box. Such an interface was built and was connected between the computer and the terminal.

In trivial terms, what it did was check every signal that came along the line from the computer – if it detected a reserved character, initially the ESCAPE character, then it sent the next character along the line to the Revox microfiche system. Essentially what this responded to was a 7-bit pattern and thus, for example, if we wanted at a particular point in the program to project frame B7 then this address would be converted to a single character, CHR\$(N), using the formula:

$$N = Z1 * 16 + Y - 1$$

where Z1 refers to A, B, C etc. and takes values of 1, 2, 3 respectively whilst Y represents the number part of the address, in this case 7. What this means is that the address B7 is represented by a single unique character string, CHR\$(38). Converting this to a binary 7-bit pattern we have 0100110 – where this bit-pattern mimics the signal that the control box would normally send out and so the microfiche is advanced to the appropriate frame.

The net result is that we now have a complete system (as shown in Figure 9.11) which, if visual material is required, is then automatically projected via the software so the user need only worry about responding to the test items raised.

At the present time this program has not been formally validated but we are hopeful that it will prove to be a very effective way of improving the quality and potential of CAL. We recently transferred the whole system onto a stand-alone microcomputer (Cromemco System Three) and the interface worked perfectly well without modification. However, with this system, to use the graphics facilities and/or the video attributes (reverse video, blinking, half-intensity etc.) does require the use of the ESCAPE character. This being the case, the nature of the reserved character was simply changed from ESCAPE to another character (Bryce and Stewart, 1980). We are also at present investigating the potential of interfacing random-access audio and also random-access TV and film and, at this early stage in the work, we are confident that it will prove of significant benefit to the students concerned.

Computer-managed Learning

Assessment, as we have seen, is an essential component of the learning process and has been dealt with in some detail in Chapter 5. Computer technology, as has been demonstrated in a number of ways, can be of considerable assistance in the management of the assessment process. For example, a number of biochemistry lecturers in the last decade have been motivated to use multiple-choice questions in the formal assessment of students. The advantages in their use are objectivity, provision of personal/individual feedback, speed of marking, wider coverage of material and promotion of self-learning (American Chemical Society, 1972, 1979; Bachman, 1975; Haywood and Wood, 1977; Bryce, 1979a; Cooper and Lockwood, 1979; Morgan, 1979a, b). Thus, multiple-choice question banks can be readily stored on computer memory, they can be accessed in a number of different ways, the tests derived from them can be rapidly marked and output in tabular form, histograms, absolute scores, relative scores and so on.

It is worth pointing out that yet another application of CML in biochemistry is the day-to-day, term-to-term administrative duties like staff and student timetabling, maintaining student and staff records, revised syllabi, standard letters or forms etc. and these can also benefit from the use of digital computers especially if some word-processing software is available, albeit primitive. The use of a computer or microprocessor can greatly facilitate the production of high-quality artwork and the generation of animated sequences which can then be used in producing learning packages in a variety of other media.

Concluding Remarks

For some lecturers the use of digital computers in education is, in some respects, a threatening and daunting development. This apprehension is largely a result of a minimal exposure to computers and computer programming and for many can be quickly overcome by attendance at a workshop on computers in education or a short programming course in, for example, BASIC where, from my own experience of running

Figure 9.11: Random-access Microfiche System Interfaced to a Main-frame Computer System and a Stand-alone Microcomputer System

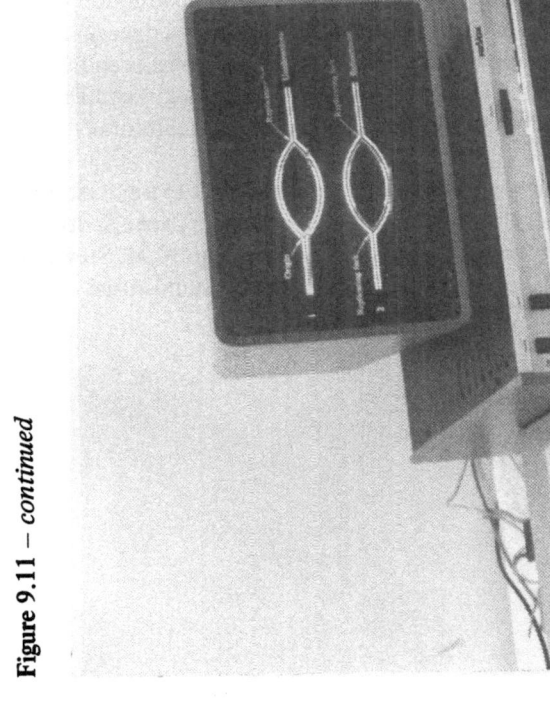

Figure 9.11 – *continued*

such a course, the participants can be writing short CAL-packages within an hour or two. Thereafter, experience and confidence can only be gained by frequent use in producing specific packages. This process can also be greatly facilitated if there is a reasonably well-established number of programs available in the particular subject area so that the newcomer can see what can be done and also provide the confidence in operating, at least, the hardware side of things. In biochemistry we *are* at such a stage of development and, provided lecturers can get ready access to a computer terminal or a stand-alone microcomputer, then there should be no reason why full use of this medium's potential cannot be accessed. Indeed, the way things are developing at present, it would be a very serious oversight for a department to disregard totally the rapid progress that is being made in this area.

Acknowledgements

I would like to thank Mr A. M. Stewart and Miss M. Gunn for their continued and active interest in the work described in this chapter and to Mr D. E. Ord for invaluable help and advice in the construction of the computer interface. I would also like to thank all my colleagues who have helped in the implementation and validation phases and in particular Dr David Button (Dundee College of Technology) and Dr Roger Griffiths (St Andrews University) for many helpful hours of discussion.

I would like to acknowledge financial support from (1) the Research Committee of the College over the last three years and (2) the Scottish Education Department for a research grant (jointly with A. M. Stewart) to evaluate the potential of interfaced random-access audio-visual equipment to improving CAL.

BIBLIOGRAPHY

AAMC (1975–6) *Survey of the Use of Computers in Instruction* (American Association of Medical Colleges, Department of Academic Affairs, Washington)

Abercrombie, M. L. J. (1960, 1969) *The Anatomy of Judgement* (Penguin, Harmondsworth)

——— (1979) *Aims and Techniques of Group Teaching*, 4th edn (Society for Research into Higher Education, Guildford)

Abercrombie, M. L. J. and Terry, P. M. (1978) *Talking to Learn: Improving Teaching and Learning in Small Groups* (Society for Research into Higher Education, Guildford)

Acheson, G. H. (1954) in Comroe, J. H. (ed.) 'The Teaching of Physiology, Biochemistry and Pharmacology', *J. Med. Educ.*, **29**(7), 20 (Suppl.)

Alexander, R. A. (1973) *J. Medical Education*, **48**, 773

American Chemical Society (1972, 1979) *National Chemistry Examination Program* (Examinations Committee, University of South Florida, Tampa, Florida)

Anderson, J. and Graham, A. (1980) *Medical Education*, **14**, 4

Anderson, R. C. and Ausubel, D. P. (1965) *Readings in the Psychology of Cognition* (Holt, Rinehart and Winston, New York)

Andriessen, J. J. and Kroon, D. J. (1980) *Educational Technology*, **20**, 21

Appleton, D. R. (1973) *American J. Physical Anthropology*, **39**, 267

Argyle, M. (1967) *The Psychology of Interpersonal Behaviour* (Penguin, Harmondsworth)

Armstrong, R. H. and Taylor, J. L. (eds.) (1970) *Instructional Simulation Systems in Higher Education* (Cambridge Monographs on Teaching Methods No. 2, Cambridge Institute of Education)

——— (1971) *Feedback on Instructional Simulation Systems* (Cambridge Institute of Education)

Ashby, E. and Besse, R. M. (1972) *The Fourth Revolution. Instructional Technology in Higher Education* (McGraw-Hill Book Company, New York)

Ashworth, J. M. (1972) *J. Biological Education*, **6**, 306

Astrup, P. (1975) *Clinical Chemistry*, **21**, 1709

Atkins, G. L. and Nimmo, I. A. (1973) *Biochemical J.*, **135**, 779

Atkinson, R. C. and Wilson, H. A. (1968) *Science*, **162**, 73

Atthill, C. (1975) *'West Oil Distribution'. A Decisions Game* (Bath University School of Education and BP Educational Service)

Ausubel, D. P. (1963) *The Psychology of Meaningful Verbal Learning* (Grune and Stratton, New York)

Ayscough, P. B. (1973) in Billing, D. E. and Furniss, B. S. (eds.) *Aims, Methods and Assessment in Advanced Science Education* (Heyden, London)

Bachman, K. A. (1975) *J. College Science Teachers*, **5**, 124

Baggott, J. (1976) *Biochemical Education*, **4**, 41

Baggot, J., Lawrence, D. M., Shaw, F., Galey, M. and Devlin, T. M. (1977) *J. Medical Education*, **52**, 157

Baggot, J. and Trojak, J. E. (1978) *Biochemical Education*, **6**, 38

Barnett, J. T. (1976) *Basic Immunology and its Medical Application* (The C. V. Mosby Co., St. Louis)

Batlle, A. M. del C. (1975) *Biochemical Education*, **3**, 28

Beard, L. (1976) *Teaching and Learning in Higher Education*, 2nd edn (Penguin, Harmondsworth)

Beard, R. M. (1970) in Armstrong, R. H. and Taylor, J. L. (eds.) *Instructional Simulation Systems in Higher Education* (Cambridge Institute of Education)
———— (1973) in Billing, D. E. and Furniss, B. S. (eds.) *Aims, Methods and Assessment in Advanced Science Education* (Heyden, London), 3

Beard, R. M., Bligh, D. A. and Harding, A. G. (1978) *Research into Teaching Methods in Higher Education*, 4th edn (Society for Research into Higher Education, Guildford)

Beard, R. M. and Pole, K. (1971) *Brit. J. Medical Education*, **5**, 13

Becher, R. A., Hewton, E., Simons, H. and Squires, G. (1973) 'Group for Research and Innovation in Higher Education', *Nuffield Foundation Newsletter*, **3**, 25

Beech, G. (1975) *FORTRAN IV in Chemistry* (Wiley, New York)

Bender, D. A. (1977) *Biochemical Education*, **5**, 49

Billing, D. E. (1973) in Billing, D. E. and Furniss, B. S. (eds.) *Aims, Methods and Assessment in Advanced Science Education* (Heyden, London), Chapter 16

Black, L. W. and Hogness, D. S. (1969) *J. Biological Chemistry*, **244**, 1976

Blaisdell, F. J. (1976) *J. Educational Technology Systems*, **5**, 155

Blanchaer, M. C. (1975) *Biochemical Education*, **3**, 71

Bligh, D. A. (1972) *What's the Use of Lectures?* (Penguin, Harmondsworth)

Blitz, A. N. (1972) Dissertation (University of Kentucky, Kentucky)

Block, G. A. (1968) National Bureau of Standards, Studies 1 and 2 (Washington, D.C.)

Bloom, B. S. (1956) *Taxonomy of Educational Objectives* (D. McKay, New York)

Bloomer, J. (1973) *Programmed Learning and Educational Technology* (Kogan Page, London)

Bloomfield, D. K., Gamble, T. E., Sorlie, W. and Andersen, J. D. (1973) *Illinois Medical Journal*, **143**, 32

Blunt, M. J. (1976) *A New Approach to Teaching and Learning Anatomy* (Butterworths, London)

Bohinski, R. C. (1979) *Modern Concepts in Biochemistry*, 3rd edn (Allyn and Bacon Inc., Boston, Mass.)

Boocock, S. and Schild, E. P. (1968) *Simulation Games in Learning* (Sage Publications, Beverley Hills, California)

Boud, D. J., Bridge, W. A. and Willoughby, L. (1975) 'P.S.I. Now – A Review of Progress and Problems', *Brit. J. Educational Technology*, **6**, 15

Boud, D. J., Dunn, J. G., Kennedy, T. and Walker, M. G. (1978) *Laboratory Teaching in Tertiary Science: A Review of Some Recent Developments* (Higher Education Research and Development Society of Australasia, Sydney)

Bramley, W. (1979) *Group Tutoring: Concepts and Case Studies* (Kogan Page, London)

Brandt, R. S. (1980) *Instructional Innovator*, **25**, 10

Branson, R. K. and Foster, R. W. (1980) *J. Educational Technology Systems*, **8**, 241

Brewer, I. M. (1974) *J. Biological Education*, 8, 101
———— (1977) *Studies in Higher Education*, 2, 33
Broderick, W. R. and Lovatt, K. F. (1975) *Brit. J. Educational Technology*, 2, 6
Broderick, W. R. and Turner, P. L. (1975) 'Computing in Schools', Fifth Int.
  Conf. (Imperial College of Science and Technology, London)
Broudy, H. (1970) *Educational Theory*, 20, 3
Bruner, J. S. (1966) *Towards a Theory of Instruction* (Harvard University Press,
  Cambridge, Mass.)
Bryce, C. F. A. (1975) Unpublished material, sample booklets available on request
———— (1977) *J. Biological Education*, 11, 140
———— (1978) *J. Biological Education*, 12, 133
———— (1979a) *Biochemical Education*, 7, 17
———— (1979b) *Trends in Biochemical Sciences*, 4, N62
———— (1979c) *Laboratory Practice*, 28, 403
———— (1979d) Fifth International Conference on Improving University Teaching,
  Course Development and Implementation, I (London), 1078
———— (1979e) *J. Biological Education*, 13, 330
———— (1980a) Nuffield Foundation Booklet (in preparation)
———— (1981a) Manuscript in preparation
———— (1981b) Manuscript in preparation
Bryce, C. F. A. and Crichton, R. R. (1971a) *J. Biological Chemistry*, 246, 4198
———— (1971b) *J. Chromatography*, 63, 267
Bryce, C. F. A. and Stewart, A. M. (1978) in Brook, D. and Race, P. (eds.)
  Aspects of Educational Technology, Vol. XII (Kogan Page, London), 256
———— (1979) in Page, G. T. and Whitlock, Q. A. (eds.) *Aspects of Educational
  Technology, Vol. XIII* (Kogan Page, London), 245
———— (1980) Institute of Medical and Biological Illustration, Second Interna-
  tional Conference (Stirling, Scotland)
Buckley-Sharp, M. D. and Harris, F. T. C. (1970) *British J. Medical Education*, 4,
  42
Buros, O. K. (1938, 1941, 1949, 1953, 1959, 1965, 1972) *The Mental
  Measurements Year Book* (Gryphon Press, Highland Park, New Jersey)
Burt, C. (1959) *A Psychological Study of Typography* (Cambridge University
  Press, London)
Bushnell, D. D. and Allen, D. W. (1967) *The Computer in American Education*
  (Wiley, New York)
Calder, J. R. (1980) *Educational Technology*, 20, 21
Callely, A. G. and Hughes, D. E. (1972) *J. Biological Education*, 6, 311
Calvo, J. (1978) *Biochemical Education*, 6, 78
Campbell, P. N. (1974) *Biochemical Education*, 2, 25
———— (1975a) *Biochemical Education*, 3, 3
———— (1975b) *Biochemical Education*, 3, 27
Candlish, J. K. (1974) *Biochemical Education*, 2, 44
Carpenter, C. R. (1971) *British J. Educational Technology*, 2, 229
Carrington, J. M. (1978) *Biochemical Education*, 6, 80
Carroll, J. B. (1974) in Olsen, D. E. (ed.) *Media and Symbols* (National Society
  for the Study of Education and University of Chicago Press, Chicago)
Carter, L. F. (1969) *Educational Technology*, 9, 22
Chapman, C. B. (1979) *Science*, 205, 559

Charren, P. (1980) *Instructional Innovator*, **25**, 10

Cheek, V. P. (1977) *Programmed Learning and Educational Technology*, **14**, 223

Chez, R. A. and O'Gorman, D. (1970) *J. Medical Education*, **45**, 807

Churchman, C. W. (1968) *The Systems Approach* (Dell Publishers, New York)

Clarke, R. (1978) *Medical J. Australia*, **1**, 255

Cleland, W. W. (1963) *Nature*, **198**, 463

Cnop-Grandsard, F. (1979) Fifth International Conference on Improving University Teaching (London), 1174

Cody, J. T. and Treagust, D. F. (1977) *Science Education*, **61**, 331

Cohen, A. J., Slovin, D. L., Franzblau, C. and Sinex, F. M. (1973) *J. Medical Education*, **48**, 289

Collier, K. G. (1966) *University Quart.*, **20**, 336

————— (1969) *University Quart.*, **23**, 431

Cooper, A. and Lockwood, F. (1979) in Page, G. T. and Whitlock, Q. A. (eds.) *Aspects of Educational Technology, Vol. XIII* (Kogan Page, London), 252

Cornish-Bowden, A. and Eisenthal, R. (1974) *Biochemical J.*, **139**, 721

Corrigan, R. E. (1980) *Educational Technology*, **20**, 26

Costin, F., Greenough, W. T. and Menges, R. J. (1972) *Review of Educational Research*, **41**, 511

Coughlan, M. P. (1978) *Biochemical Education*, **6**, 16

Coulson, J. E. (1962) *Programmed Learning and Computer-Based Instruction* (Wiley, New York)

Cowan, J. (1975) in Gibbs, G. I. and Howe, A. (eds.) *Academic Gaming and Simulation in Education and Training* (PIC/AGSET Publication, Kogan Page, London)

————— (1980) *Programmed Learning and Educational Technology*, **17**, 115

Cronin, N. R. (1979) Fifth International Conference on Improving University Teaching (London), 904

Crosby, J. L. (1961) *Heredity*, **16**, 255

Cross, K. P. (1976) *Accent on Learning* (Jossey-Bass, London)

Crovello, T. J. (1974) *Bioscience*, **24**, 20

Culter, P. (1979) *Problem Solving in Clinical Medicine* (Williams and Wilkins, Baltimore)

Cunningham, P. (1979) *Biochemical Education*, **7**, 83

Cuypers, P. A. (1978) *Biochemical Education*, **6**, 76

Dakshinamurti, K. (1979) *Biochemical Education*, **7**, 70

Dalziel, K. (1979) *Trends in Biochemical Sciences*, **4**, 231

Daubert, S. D. and Sontum, S. F. (1977) *J. Chemical Education*, **54**, 35

Davies, I. K. (1971) *The Management of Learning* (McGraw-Hill, London)

————— (1976) *Objectives in Curriculum Design* (McGraw-Hill, London)

Davies, M. A., Gale, J. and Clarke, W. D. (1977) *Medical Education*, **11**, 370

Davies, P. L. (1980) *Biochemistry, Level III* (MacDonald and Evans, Plymouth)

Davis, R. H., Alexander, L. T. and Yelon, S. L. (1974) *Learning System Design* (McGraw-Hill, New York)

de Bono, E. (1970) *Lateral Thinking* (Penguin, Harmondsworth), 243

de Winter Hebron, C. C. (1979) *Bulletin of Educational Research*, **17**, 24

Dee, J. (1976) *Group Discussions in Biological Sciences*, The Nuffield Foundation, Group for Research and Innovation in Higher Education

Deland, E. C. (1978) *Information Technology in Health Sciences Education*

(Plenum Publishing Corp., New York)

Desrosier, E. V. (1976) PhD Dissertation (Boston University School of Education)

Devlin, T. M. (1979) Personal communication to F. Vella

Dimitriou, B. (1971) in Armstrong, R. H. and Taylor, J. L. (eds.) *Feedback on Instructional Simulation Systems* (Cambridge Institute of Education), 45

Dixon, J. E. (1978) *Biochemical Education*, 6, 3

Dodge, M., Bogdan, R., Brodgen, N. and Lewis, R. (1974) *Educational Technology*, 14, 21

Dowdeswell, W. H. (1973) *J. Biologica' Education*, 7, 8

―――― (1974) *May and Baker Laboratory Bulletin*, 11, 6

Duchastel, P. C. (1978) *Educational Technology*, 11, 36

Duchastel, P. C. and Whitehead, D. (1980) *Programmed Learning and Educational Technology*, 17, 41

Dwyer, F. (1972) *Improving Visualized Instruction* (State College, Pennsylvania)

Eadie, G. S. (1942) *J. Biological Chemistry*, 146, 85

Editorial (1970) *Lancet*, 1, 984

Eggen, P. D., Kauchak, D. P. and Harder, R. J. (1979) *Strategies for Teachers* (Prentice-Hall, Englewood Cliffs, New Jersey)

Elton, L. R. B., Boud, D. J., Nuttall, J. and Stace, B. C. (1973) *Chemistry in Britain*, 9, 164

Epstein, H. T. (1970) *A Strategy for Education* (Oxford University Press, Oxford)

Essenberg, R. C. (1979) *Biochemical Education*, 7, 85

―――― (1980) *LOADCO and COKIN, Computer Kinetics Project* (Oklahoma State University Press, Oklahoma)

Evered, D. F. (1975) *Biochemical Education*, 2, 53

Farnsworth, T. (1979) *Biochemical Education*, 7, 63

Fedoroff, S. and Opel, W. (1978) *J. Medical Education*, 53, 415

Fleming, J. and Stuckey, J. (1972) 'Explorations in Small Group Learning' (Research and Development Paper no. 16, Tertiary Education Research Centre, University of New South Wales, Australia)

Fogel, B. J. (1976) *J. Nat. Medical Association*, 68, 170

Folk, R. L., Griesen, J. V. and Beran, R. L. (1976) *Individualising the Study of Medicine* (The Ohio State University College of Medicine Independent Study Program, Westinghouse Learning Corp., Illinois), 1

Fox, J. L. (1978) *Chemical and Engineering News*, 56, 16

Fox, R. G. (1979) *J. Educational Technology Systems*, 7, 229

France, V. M. (1979) Cited in *The Physiologist*, 22, 27

Freitag, B. (1980) *Instructional Innovator*, 25, 10

Fresco, J. R., Alberts, B. M. and Doty, P. (1960) *Nature*, 188, 98

Friedland, J. and Rosen, B. (1971) *LOCKEY, The Lock and Key Model of Enzyme Action* (Huntingdon Two Computer Project, Digital Equipment Corporation, New York)

Frunder, H. E. (1978) *Biochemical Education*, 6, 75

Fukagawa, Y., Takei, T. and Ishikawa, T. (1980) *Biochemical J.*, 185, 177

Gagné, R. M. (1962) *Psychological Principles in System Development* (Holt, Rinehart and Winston, New York)

―――― (1963) *J. Res. Science Teaching*, 1, 144

Gagné, R. M. and Briggs, L. J. (1974) *Principles of Instructional Design* (Holt, Rinehart and Winston, New York)

Galindo, J. D., Pedreno, E., Canovas, F. G. and Lozano, J. A. (1976) *Revista de Educacion*, **244**, 112

Garcia, E. N., Hull, E. W., Ibsen, K. H. and Paseman, F. H. (1968) in Lysaught, J. P., Sutherland, S. A. and Mullen, P. A. (eds.) *Individualised Instruction in Medical Education* (The Rochester Clearinghouse on Self-Instructional Materials for Health Care Facilities, University of Rochester), 74

Garland, P. B. (1975) *Biochemical Education*, **3**, 42

Garland, P. B., Dutton, G. J. and MacQueen, D. (1977) *Studies in Higher Education*, **2**, 167

Garvey, D. M. (1971) in Tansey, P. J. (ed.) *Educational Aspects of Simulation* (McGraw-Hill, London), 204

Gayford, D. M. (1979) *School Science Review* (September), 180

Gentile, J. R. (1967) *AV Communication Review*, **15**, 25

Gerlach, V. S. and Ely, D. P. (1971) *Teaching and Media, a Systematic Approach* (Prentice-Hall, Eaglewood Cliffs, New Jersey)

Gerrick, D. J. (1971) *The American Biology Teacher*, **33**, 526

Gibbs, G. I. (1975) in Gibbs, G. I. and Howe, A. (eds.) *Academic Gaming and Simulation in Education and Training* (PIC/AGSET Publication, Kogan Page, London), 7

Glaser, R. (1965) 'Teaching Machines and Programmed Learning, II, Data and Directions' (Dept. of Audio-Visual Instruction of the National Education Association, Washington)

—— (1968) *Proc. of the 1967 Invitational Conference on Testing Problems* (Educational Testing Service, Princeton)

Goldberg, J. L., Houghton, B. J. and Kind, P. R. W. (1970) *Brit. J. Medical Education*, **4**, 117

Goldstein, J. P. and Brown, J. S. (1979) in Lockhead, J. and Clement, J. (eds.) *Cognitive Process Instruction* (The Franklin Institute Press, Philadelphia), 201

Gray, C. H. (1975) *Biochemical Education*, **3**, 24

Gray, F. D. and Soffer, A. (1980) 'Internal Medicine', *J. American Medical Association*, **243**, 2190

Grimm, F. M. (1978) *The American Biology Teacher*, **40**, 362

Grundin, H. V. (1969) *Aspects of Educational Technology, Vol. III* (Pitman, London), 65

Gruneberg, M. M. and Startup, R. (1978) *The Vocational Aspect of Education*, **30**, 75

Guyer, K. E., Poland, J. L. and Seibel, H. R. (1975) *Southern Medical J.*, **68**, 1120

Halcomb, J. D. and Garner, A. E. (1973) *Improving Teaching in Medical Schools (A Practical Handbook)* (Charles C. Thomas, Springfield, Illinois), 92

Hall, J. (1973) *California Management Journal*, **15**, 56

Hall, P. F. (1973) *Medical J. Australia*, **2**, 153

Hammond, R. A. and Roach, D. K. (1976) *Medical and Biological Illustration*, **26**, 87

Hancock, V. J. F. (1978) *J. Biological Education*, **12**, 284

Harden, R. McG. (1979) *Medical Teacher*, **1**, 289

Harden, R. McG., Dunn, W. R., Murray, T. S., Rogers, J. and Stoane, C. (1979) *Brit. Medical J.*, **2**, 652

Harden, R. McG., Lever, R., Dunn, W. R., Lindsay, A., Holroyd, C. and Wilson, G. M. (1969) *Lancet*, **1**, 933

Harding, R. D. (1980) *Studies in Higher Education*, 5, 101

Harper, J. L. (1972) *J. Biological Education*, 6, 318

Harpp, D. N. and Snyder, J. P. (1977) *J. Chemical Education*, 54, 68

Harrow, A. J. (1972) *A Taxonomy of the Psychomotor Domain* (D. McKay, New York)

Hartley, J. (1978) *Designing Instructional Text* (Kogan Page, London)

Hartley, J. and Burnhill, P. (1976) *Bulletin of the British Psychological Society*, 29, 97

—— (1978) in Hartley, J. and Davies, I. K. (eds.) *Contributions to an Educational Technology*, Vol. 2 (Kogan Page, London: Nichols, New York)

Hartley, J. and Davies, I. K. (1976) *Review of Educational Research*, 46, 239

—— (1978) *Programmed Learning and Educational Technology*, 15, 207

Hartley, J. and Mills, R. L. (1973) *Brit. J. Educational Technology*, 4, 120

Hartley, J. R. and Lovell, K. (1977) in Jones, A. and Weinstock, H. (eds.) *Computer-Based Science Instruction* (Noordhoff International, Groningen)

Hartman, J. B. (1978) *CAUT Bulletin* (May), 19

Haywood, J. and Wood, E. J. (1977) *Biochemical Education*, 5, 36

Heidt, E. V. (1975) *British J. Educational Technology*, 6, 4

Herskovitz, A. (1979) *J. Biocommunication*, 6, 19

Hertzler, E. C. (1979) *Educational Technology*, 19, 45

Heydeman, M. T. (1977) *ENZKIN, Unit on Enzyme Kinetics* (Chelsea Science Simulation Project; Edward Arnold, London)

Hill, W. F. (1962, 1977) *Learning Thru Discussion* (Sage Publications Inc., Beverley Hills)

Hofstee, B. H. J. (1952) *Science*, 116, 329

Hooper, R. (1977) *Computer-Assisted Learning in Higher Education: The Next Ten Years* (Council for Educational Technology, London)

—— (1978) *Computers and Education*, 2, 1

Hooper, R. and Toye, I. (1975) *Computer-Assisted Learning in the United Kingdom: Some Case Studies* (Council for Educational Technology, London)

Horn, R. E. (1977) *The Guide to Simulation/Games for Education and Training*, Vol. 1, 3rd edn (Didactic Systems Inc., Cranford, New Jersey)

Horsey, J. and Milson, A. (1980) *Programmed Learning and Educational Technology*, 17, 36

Howe, M. J. A. and Godfrey, J. (1978) *Student Note-Taking as an Aid to Learning* (Exeter University Teaching Service, Exeter)

Huey, E. B. (1908, 1968) *The Psychology and Pedagogy of Reading* (Macmillan, London, 1908; MIT Press, Cambridge, Mass., 1968)

Hull, P. (1977) *J. Biological Education*, 11, 202

Hull, P. (1978) *J. Biological Education*, 12, 21

Hultquist, D. E. (1976) *Biochemical Education*, 4, 3

Hulquist, D. E., Marino, J. P., Martin, M. M. and Price, A. R. (1976) *J. Medical Education*, 51, 57

Hunt, G. J. F. (1977) *Programmed Learning and Educational Technology*, 14, 74

Huntington, J. F. (1979) *Computer-Assisted Instruction Using BASIC* (Educational Technology Publications, Englewood Cliffs, New Jersey)

Huxham, G. T. and Naeraa, N. (1980) *Medical Education*, 14, 23

Hyatt, G. W., Eades, D. C. and Tenczer, P. (1972) *Bioscience*, 22, 401

Hyde, R., Lanier, R. A., Billington, D. R. and Sengel, R. A. (1976) Research

Report No. 8 (Learning Resources Center, Oklahoma Health Sciences Center)

Hyzer, W. G. (1977) *Research Development*, **28**, 20

Iborra, J. L. and Lozano, J. A. (1980) *Biochemical Education*, **8**, 57

Jacobson, E. J. (1971) *Gastroenterology*, **60**, 955

Jepson, J. B. and Smith, A. D. (1972) in Austwick, K. and Harris N. D. C. (eds.) *Aspects of Educational Technology*, Vol. VI (Pitman Publishers, London)

Jepson, J. B. and Smith, A. D. (1973) in Hill, P. J., Palmer, C. R. and Trickey, D. A. (eds.) *Alternatives to the Lecture in Chemistry* (The Chemical Society, London), 39

Jones, T. H. D. (1976) *J. College Science Teaching*, **5**, 316

Josse, J., Kaisser, A. D. and Kornberg, A. (1961) *J. Biol. Chem.*, **236**, 864

Joyce, B. and Weil, M. (1972) *Models of Teaching* (Prentice-Hall, Englewood Cliffs, New Jersey)

Judisch, J. M. (1968) *Nat. Micrographics Association J.*, **2**, 58

Karlson, P. (1977) *Biochemical Education*, **5**, 1

Keller, F. S. (1968) *J. Applied Behavioural Analysis*, **1**, 78

Kellerman, G. M. (1978) *Biochemical Education*, **6**, 76

Kerr, A. M. (1970) *U. Manitoba Medical Journal*, **42**, 25

Kidd, N. A. C. (1979) *J. Biological Education*, **13**, 284

Klein, J. (1961) *Working with Groups* (Hutchinson, London)

Kogut, M. (1974) *Biochemical Education*, **2**, 51

———— (1976) *Biochemical Education*, **4**, 9

Kogut, M. and Cramp, D. (1975) *Brit. J. Medical Education*, **9**, 188

Kolers, P. A. (1968) *Perception and Psychophysics*, **3**, 57

———— (1969) *Journal of Typographical Research*, **3**, 145

Kornberg, A. (1961) *The Enzymatic Synthesis of DNA* (Wiley, London)

Kornberg, A. (1978) *Trends in Biochemical Sciences*, **34**, N73

Kozma, R. B., Belle, L. W. and Williams, G. W. (1978) *Instructional Techniques in Higher Education* (Educational Technology Publications, Englewood Cliffs, New Jersey)

Krathwohl, D. R., Bloom, B. S. and Masia, B. B. (1964) *Taxonomy of Educational Objectives. Handbook II: Affective Domain* (D. McKay, New York)

Kuhn, E. and Brand, K. (1973) *Biochemistry*, **12**, 5217

Kulik, J. A., Kulik, C.-L. and Carmichael, K. (1974) *Science*, **183**, 379

Laburn, H. B., Mendelow, A. D., Cantrill, R. and McCalden, T. (1978) Cited in *South African Med. J.*, **54**, 885

Landon, M. and Mayer, R. J. (1976) *Biochemical Education*, **4**, 22

Leader, D. P. (1976) *J. Biological Education*, **10**, 303

Lee, A. (1978) *Medical Journal of Australia*, **1**, 497, 551, 605, 645

Lee, D. R. and Buck, J. R. (1975) *Human Factors*, **17**, 461

Lehninger, A. L. (1975) *Biochemistry*, 2nd edn (Worth Publishers Inc., New York)

Leiblum, M. (1977) *UCODI Bulletin*, **13**, 1

Levie, W. H. and Dickie, K. E. (1973) in Travers, R. (ed.) *Second Handbook of Research on Training* (Rand McNally, Chicago, Ill.)

Lineweaver, H. and Burk, D. (1934) *J. American Chemical Society*, **56**, 658

Long, B. W. (1979) *The American Biology Teacher*, **41**, 112

MacDonald-Ross, M. (1972) in Austwick, K. and Harris, N. D. C. (eds.) *Aspects of Educational Technology, Vol. VI* (Pitman, London)

———— (1973) *Instructional Science*, **2**, 1
———— (1978) *Review of Research in Education*, **6**, 229
MacQueen, D., Chignell, D. A., Dutton, G. J. and Garland, P. B. (1976) *Medical Education*, **10**, 418
MacQuire, C. H. and Solomon, L. M. (1971) *Clinical Simulations – Selected Problems in Patient Management* (Meredith Corp., New York)
Maffet, J. E. (1967) *Instructional Performance Objectives for a Course in General Biology (Manatee Junior College)* (University of California Clearinghouse, Los Angeles)
Magar, R. F. (1962, 1975) *Preparing Instructional Objectives* (Fearon Publishers, California)
———— (1968) *Developing Attitude Toward Learning* (Fearon Publishers, California)
Magne, O. and Parknas, L. (1963) *Brit. J. Educational Psychology*, **33**, 265
Marmion, B. P. and Lutz, W. (1971) in Whitby, L. G. and Lutz, W. (eds.) *Principles and Practice of Medical Computing* (Churchill Livingstone, Edinburgh)
Mason, J. H. (1979) *Visual Education* (February), 29
McAshen, H. H. (1970) *Writing Behavioural Objectives: A New Approach* (Harper and Row, London)
McBrien, D. (1980) *Biologist*, **27**, 57
McIntyre, D. N. (1975) *Biochemical Education*, **3**, 6
McKenzie, J., Elton, L. R. B. and Lewis, R. (1978) *Interactive Computer Graphics in Science Teaching* (Ellis Horwood, Chichester)
Mehler, A. H. (1973) *Biochemical Education*, **1**, 26
Michaelis, L. and Menten, M. L. (1913) *Biochem. Z.*, **49**, 333
Miller, A. L., Mills, G. L. and Smith, A. D. (1979) *Biochemical Education*, **7**, 31
Miller, L. D., Winegard, S. S., Brooks, F. P., Fox, R. C. and Rosato, F. E. (1972) *Surgery*, **72**, 323
Miller, M. (1979) Fifth International Conference on Improving University Teaching (London), 1204
Mitzel, H. E. (1966) American Management Association's 2nd Int. Conf. (New York)
Monro, D. M. (1978) *Basic BASIC* (Edward Arnold, London)
Montgomery, R., Dryer, R. L., Conway, T. A. and Spector, A. A. (1974, 1977, 1980) *Biochemistry – A Case-Oriented Approach* (The C. V. Mosby Co., St Louis, Missouri)
Montgomery, R. and Swenson, C. A. (1976) *Quantitative Problems in Biochemical Sciences*, 2nd edn (W. H. Freeman and Co., San Francisco)
Moore, J. W. and Collins, R. W. (1979) *J. Chemical Education*, **56**, 140
Moore, J. W., Gerhold, G., Breneman, G. L., Owen, G. S., Butler, W., Smith, S. G. and Lyndrup, M. L. (1979) *J. Chemical Education*, **56**, 776
Morgan, M. R. J. (1979a) *Biochemical Education*, **7**, 67
———— (1979b) *Biochemical Education*, **7**, 84
Muir, F. (1978) *The Frank Muir Book: An Irreverent Companion to Social History* (Corgi Books, London)
Murray, W. C. (1979) *Biochemical Education*, **7**, 77
Nelson, C. E. (1965) *Microfilm Technology* (McGraw-Hill, New York)
Neufeld, V. R. (1974) in Bowers, J. Z. and Purcell, E. F. (eds.) *Teaching the Basic

Medical Sciences: Human Biology (Josiah Macy, Jr. Foundation, New York), 122

Neufeld, V. R. and Barrows, H. S. (1974) *J. Medical Education*, **49**, 1040

Newsholme, E. A. (1979) *Trends in Biochemical Sciences*, **4**, N188

Nimmo, I. A. and Flynn, I. W. (1978) *Biochemical Education*, **6**, 42

Nyman, P. and Lindskog, D. S. (1964) *Biochem. Biophys. Acta*, **85**, 141

Ogborn, J. (1977) *Small Group Teaching in Undergraduate Science* (Heinemann Educational, London)

Orr, H., Marshall, J. C., Isenhour, T. L. and Jurs, P. C. (1973) *Introduction to Computer Programming for Biological Scientists* (Allyn and Bacon Inc., Boston)

Ottaway, A. K. C. (1968) in Layton, D. (ed.) *University Teaching in Transition* (Oliver and Boyd, Edinburgh), 53

Owen, S. G., Hall, R., Anderson, J. and Smart, G. A. (1965) *Postgraduate Medical J.*, **41**, 201

Parsons, H. M. (1974) *Science*, **183**, 922

Pasternak, C. A. (1979) *An Introduction to Human Biochemistry* (Oxford University Press, Oxford)

Paterson, D. G. and Tinker, M. A. (1932) *J. Applied Psychology*, **16**, 605

Peacock, D. and Tribe, M. A. (1979) *The Enzyme Game* (Cambridge University Press, Cambridge)

Pearlmutter, A. F. and Pearlmutter, F. A. (1977) *Biochemical Education*, **5**, 5

Pinnick, H. R. Jr. (1980) *Quantum Chemistry Program Exchange*, **12**, 384

Popham, W. J. and Baker, E. L. (1970) *Systematic Instruction* (Prentice-Hall, Englewood Cliffs, New Jersey)

Postlethwait, S. N., Novak, J. and Murray, H. T. (1971) *The Audio-Tutorial Approach to Learning*, 2nd edn (Burgess Publishing Company, Minneapolis, Minnesota)

Poulton, E. C. (1972) in Davies, I. K. and Hartley, J. (eds.) *Contributions to an Educational Technology* (Butterworth, London)

Powell, J. P. (1971) *Universities and University Education. A Selected Bibliography, Vol. 2* (National Foundation for Educational Research in England and Wales, London)

——— (1974) *Educational Research*, **16**, 163

——— (1977) *Higher Education: A Selected Bibliography, Vol. 3, 1970–5* and *Supplement to Vol. 1* (Higher Education Research and Development Society of Australasia, Sydney)

Prior, J. A., Griesen, J. V. and Folk, R. L. (1970) *J. Medical Education*, **45**, 801

Race, W. P. (1979) in Page, G. T. and Whittock, Q. A. (eds.) *Aspects of Educational Technology, Vol. XIII* (Kogan Page, London), 195

Rackham, N. (1970) in Armstrong, R. H. and Taylor, J. L. (eds.) *Instructional Simulation Systems in Higher Education* (Cambridge Institute of Education), 203

Ramsay, H. P. (1977) *J. Biological Education*, **11**, 95

Ramsey, H. P., Wood, A. E. and Anderson, J. M. E. (1978) 'Biology of Mankind: An Innovative Course for First Year Students' (Occasional Publication No. 11, Tertiary Education Research Centre, University of New South Wales, Australia)

Randle, P. J. (1975) *Biochemical Education*, **3**, 23

Rappaport, E. (1971) 'CAI Center Technical Memorandum' (Florida State University), 33

Rasmussen, R. S. (1972) 'Biochemistry, Science (Experimental)' (Dade County Public Schools, Miami, Florida)

Rezler, A. G. (1973) Public Health Paper No. 52 (World Health Organisation, Geneva), 70

Rickards, J. P. and Di Vesta, F. J. (1974) *J. Educational Psychology*, 66, 354

Roach, D. K. and Hammond, R. A. (1976) *Studies in Higher Education*, 1, 179

Robbins, J. (1972) 'Course Evaluation of General and Human Biology' (Research and Development Paper No. 18, Tertiary Education Research Centre, University of New South Wales, Australia)

Roberts, D. V. (1977) *Enzyme Kinetics* (Cambridge University Press, Cambridge)

Robinson, F. (1961) *Effective Study* (Harper and Row, New York)

Rogers, C. R. (1969) *Freedom to Learn* (Charles E. Merril, Colombus, Ohio)
—— (1970) *Encounter Groups* (Penguin, Harmondsworth)

Rothkopf, E. Z. (1976) in Gage, N. L. (ed.) *Psychology of Teaching Methods* (National Society for the Study of Education and University of Chicago Press, Chicago)

Rubin, J. (1957) *National Micro News*, 29 (January), 1

Rubinstein, D. (1972) *J. Medical Education*, 47, 198

Ruddock, J. (1978) *Learning Through Small Group Discussion* (Society for Research into Higher Education, Guildford)

Rushby, N. J. (1979) *An Introduction to Educational Computing* (Croom Helm, London)
—— (1980) *Visual Education* (March), 8

Sable, H. Z. (1975) *J. Medical Education*, 50, 296

Saffran, M. (1971) *J. Medical Education*, 46, 1080
—— (1973) *Biochemical Education*, 1, 50

Saffran, M. and Franco-Saenz, R. (1975) *Biochemical Education*, 3, 10

Schepartz, B. (1974) *Biochemical Education*, 2, 12

Schwartz, P. L. (1975) *J. Medical Education*, 50, 903
—— (1979) *Trends in Biochemical Sciences*, 4, N85
—— (1980) *Biochemical Education*, 8, 11

Shone, J. (1979) *J. Biological Education*, 13, 123

Short, A. H. (1979) *The Physiologist*, 22, 27

Sinclair, D. (1975) *Lancet*, 1, 875

Skinner, B. F. (1968) *The Technology of Teaching* (Appleton-Century-Crofts, New York)

Smaje, L. H. and Lynn, B. (1979) *Medical Education*, 13, 129

Smith, A. D. and Jepson, J. B. (1972) *Lancet*, 2, 585

Smith, K. U. (1960) *AV Communication Review*, 4, 27

Smith, S. G. and Sherwood, B. A. (1976) *Science*, 192, 344

Smythe, A. B. (1974) *Aust. Univers.*, 12, 147

Smythe, R. and Lovatt, K. F. (1979) *J. Biological Education*, 13, 207

Spencer, H. (1969) *The Visible Word* (Lund Humphries, London)

Spencer, T. (1979) *Biochemical Education*, 7, 51

Spilman, E. L. and Spilman, H. W. (1975) *J. Medical Education*, 50, 667

Stewart, A. M. (1979) Personal communication

Stohs, S. J. and Rosenberg, H. (1976) *American J. Pharmaceutical Education*, 40, 155

Stolurow, L. M. (1967) 'Computer-Assisted Instruction' (Technical Report No. 2, ONR Contract No. N00014-67-A-0298-003, Harvard Computing Center, Cambridge, Mass.)

Stolurow, L. M., Peterson, T. I. and Cunningham, A. C. (1970) *Computer Assisted Instruction in the Health Professions* (Entelek Inc., Newburyport, Mass.)

Stoward, P. J. (1976) *Medical Education*, **10**, 316

Strang, R. H. C. (1977) *Biochemical Education*, **5**, 29

Suckling, K. E., Apps, D. K., Flynn, I. W., Nimmo, I. A., Phillips, J. H. and Van Heyningen, S. (1979) *Biochemical Education*, **7**, 54

Summers, M. K. and Summers, J. M. (1976) *J. Biological Education*, **10**, 229
—————— (1977) *J. Biological Education*, **11**, 211

Suppes, P. (1967) in Bushnell, D. D. and Allen, D. W. (eds.) *The Computer in American Education* (Wiley, New York), 11

Tansey, P. J. and Unwin, D. (1969) *Simulation and Gaming in Education* (Methuen Educational Ltd, London)

Taylor, J. L. and Carter, K. R. (1971) in Armstrong, R. H. and Taylor, J. L. (eds.) *Feedback on Instructional Simulation Systems* (Cambridge Institute of Education), 59

Taylor, J. L. and Walford, R. (1978) *Learning and the Simulation Game* (Open University Educational Enterprises Ltd, Milton Keynes)

Thomas, J. A. (1976) *Educational Technology*, **16**, 34

Thomas, K. W. (1972) *J. Biological Education*, **6**, 314

Thompson Bowles, L. (1974) in Bowers, J. Z. and Purcell, E. F. (eds.) *Teaching the Basic Medical Sciences: Human Biology* (Josiah Macy Jr. Foundation, New York), 1

Tinker, M. A. (1963) *Legibility of Print* (Iowa State University Press, Ames, Iowa)

Tinker, M. A. and Paterson, D. G. (1928) *Journal of Applied Psychology*, **12**, 359

Tobias, S. (1969) *A V Communication Review*, **17**, 3

Tomlinson, R. (1976) *J. Biological Education*, **10**, 65

Tomlinson, D. R. (1979) *Medical Education*, **12**, 257

Tosteson, D. C. (1970) *J. Medical Education*, **45**, 557
—————— (1979) *New England J. Medicine*, **301**, 690

Treagust, D. F. and Cody, J. T. (1977) *The American Biology Teacher*, **39**, 556

Tribe, M. A., Eraut, M. R. and Snook, R. K. (1975) *Basic Biology Course* (Cambridge University Press, Cambridge)

Tribe, M. A. and Peacock, D. (1973) in Budgett, R. and Leedham, J. (eds.) *Aspects of Educational Technology, Vol. VII* (Pitman, London), 292
—————— (1976) *The Ecology Game* (Cambridge University Press, Cambridge)

Trzebiatowski, G. L. (1976) in Purcell, E. F. (ed.) *Recent Trends in Medical Education* (Josiah Macy Jr. Foundation, New York), 111

Tyler, R. W. and White, S. H. (1979) Conference Report (National Institute of Education, US Department of Health, Education and Welfare, October)

UTMU (1972) *Varieties of Group Discussion in University Teaching* (University Teaching Methods Unit, London)

Varley, M. E. (1972) *J. Biological Education*, **6**, 304

Vella, F. (1975) *Biochemical Education*, **3**, 8
—————— (1977) *Biochemical Education*, **5**, 65
—————— (1979a) *Biochemical Education*, **7**, 53
—————— (1979b) *Biochemical Education*, **8**, 25

Vella, F. and Martin, R. O. (1975) *Biochemical Education*, **3**, 8

Vella, F. and Martin, R. O. (1976) *Biochemical Education*, **4**, 43

Vernon, M. D. (1953) *Brit. J. Educational Psychology*, **23**, 180

Wager, W. (1980) *Educational Communication and Technology*, **28**, 19

Wallace, S. (1978) in Kozma, R. B., Belle, L. W. and Williams, G. W. (eds.) *Instructional Techniques in Higher Education* (Educational Technology Publications, Englewood Cliffs, New Jersey), Chapter 21

Waller, R. (1977) *Typographical Access Structures for Educational Texts* (Institute of Educational Technology: The Open University, Milton Keynes)

Weed, L. L. (1969) *Medical Records, Medical Education, and Patient Care* (The Press of Case Western Reserve University, Cleveland)

Weinberg, S. B. (1979) Fifth International Conf. on Improving University Teaching (London), 1875

Weisman, R. A. and Shapiro, D. M. (1973) *J. Medical Education*, **48**, 934

Whitby, L. G. (1974) *Lancet*, **1**, 683

Wieker, H.-J., Johannes, K.-J. and Hess, B. (1970) *FEBS Letters*, **8**, 178

Wight, A. R. (1972) *Educational Technology*, **12**, 9

Wildman, T. M. (1980) *Educational Technology*, **20**, 16

Wilkinson, G. L. (1976) *Communication Review*, **24**, 413

Wilkinson, G. N. (1961) *Biochemical J.*, **80**, 324

Wills, E. D. (1974) *Lancet*, **2**, 217

Winkler, B. C. (1978) *Biochemical Education*, **6**, 82

Winn, W. (1979) *J. Biocommunication*, **6**, 14

Wood, A. (1975) in Hooper, R. and Toye, I. (eds.) *Computer-Assisted Learning in the United Kingdom* (Charlesworth, Huddersfield)

Wood, A. E. (1979) *Studies in Higher Education*, **4**, 203

Woolf, B. (1932) in Haldane, J. B. S. and Stern, K. G. (eds.) *Allgemeine Chemie der Enzyme* (Steinkopff Verlag, Dresden and Leipzig), 119

Zachrisson, B. (1965) *Studies in the Legibility of Printed Text* (Almquist and Wiksell, Stockholm)

Zuckerman, D. and Horn, R. E. (1973) *The Guide to Simulations/Games for Education and Training* (Information Resources Incorporated, Lexington, Mass.)

D. Rex Billington, MA, PhD, is a Senior Lecturer in the Department of Community and Occupational Medicine of the University of Dundee where he also directs the Health Behaviour Research Unit. Dr Billington was born in New Zealand and spent ten years in academic institutions in the USA, the last part of which was in directing a university learning-resource centre. His association and interest in biochemistry education and its assessment grew from this position. He currently has an honorary appointment in the University of Dundee Centre for Medical Education. Dr Billington's qualifications are in education and the behavioural sciences which he teaches to public health, medical and dental students.

Charles F. A. Bryce is Senior Lecturer in Biochemistry at Dundee College of Technology where he has taught undergraduate biochemistry to degree (Science and Nursing) and diploma students since 1975. Prior to that he was actively involved in research at University College, Cardiff, the Max Planck Institute of Molecular Genetics, Berlin and the University of Glasgow with a research interest largely in protein chemistry and enzymology. In the period since his PhD graduation he has formally attended two postgraduate courses, one in computing and numerical methods, the other in educational technology, with the view to developing a strong base from which to study methods for improving the quality and effectiveness of his teaching. In addition to this, he has undertaken two periods of secondment to the Audio-Visual and Reprographic Unit of the college for the purpose of designing, implementing and evaluating self-instructional learning packages in biochemistry and molecular biology.

Krishnamurti Dakshinamurti has been Professor of Biochemistry at the University of Manitoba since 1973. Prior to this he had been Associate Professor (University of Manitoba), Visiting Professor (Rockefeller University), Associate Director (St Joseph Hospital), and Research Associate (Massachusetts Institute of Technology and University of Illinois). He received the Borden Award of the Nutritional Society of Canada in 1973. His research interests are in metabolic control mechanisms and maturation of the central nervous system.

Rex Montgomery is Professor of Biochemistry and Associate Dean for Academic Affairs of the College of Medicine at the University of Iowa. He was Director of the Physician's Assistant Program. Professor Montgomery's primary research contributions have been in the chemistry and biochemistry of carbohydrates, glycoproteins and the cell membrane of normal and malignant cells. He is the author of *The Chemistry of Plant Gums and Mucilages* (1959), *Quantitative Problems in the Biochemical Sciences* (with C. A. Swenson; 2nd edn, 1976) and *Biochemistry: A Case-Oriented Approach* (with R. L. Dryer, T. W. Conway, and A. A. Spector; 3rd edn, 1980). He received his BSc, PhD, and DSc degrees from the University of Birmingham, England (1943, 1946, 1963).

Alistair M. Stewart is Director of the Audio-Visual and Reprographic Unit at Dundee College of Technology. Having initially qualified in chemistry, he later transferred to educational technology, gained an MSc in chemical education, and is currently engaged in research into the design and utilisation of print as an instructional medium. Since 1976 he has regularly acted as a consultant in educational technology to the World Health Organisation, and is presently actively involved in a staff development project at the University of Cairo, Egypt.

Michael Tribe has been a Lecturer in Biology at The University of Sussex since 1966. Previous to that he was a schoolteacher for four years. He has therefore maintained a strong interest in science education for many years. In 1969 he was appointed Director of the Sussex component of the Inter-university Biology Teaching Project, resulting in the publication of a Basic Biology Course for undergraduates in 1975–8. From the period 1973–6 he was also involved in writing materials for the Joint Universities Genetics Project, published as 'Genetics' S299 by the Open University. His biological research interests have mainly been concerned with the biochemistry and genetics of mitochondria, where he has published several papers and two books, the latter in collaboration with Dr Peter Whittaker.

F. Vella graduated MD in 1952 from the Royal University of Malta and proceeded to Oxford University as Rhodes Scholar. He is currently Professor of Biochemistry at the University of Saskatchewan, Saskatoon, Canada where he went in 1965 after five years at the University of Khartoum, Sudan and prior to that, four years at the University of Malaya in Singapore. His research activities have been in the fields of haemoglobinopathies, thalassaemias and erythrocyte enzymopathies.

Alec E. Wood is a Lecturer in Botany at the University of New South Wales with his area of interest being mycology. Prior to 1980, he acted as Director of the First Year Biology Teaching Unit and as a result of this work developed a keen interest in the structuring etc. of small-group tutorials.

INDEX